DANCING WITH LIGHT

DANCING WITH LIGHT

Advances in Photofunctional
Liquid-Crystalline Materials

Haifeng Yu

PAN STANFORD PUBLISHING

Published by

Pan Stanford Publishing Pte. Ltd.
Penthouse Level, Suntec Tower 3
8 Temasek Boulevard
Singapore 038988

Email: editorial@panstanford.com
Web: www.panstanford.com

British Library Cataloguing-in-Publication Data
A catalogue record for this book is available from the British Library.

Dancing with Light: Advances in Photofunctional Liquid-Crystalline Materials

Copyright © 2015 by Pan Stanford Publishing Pte. Ltd.

All rights reserved. This book, or parts thereof, may not be reproduced in any form or by any means, electronic or mechanical, including photocopying, recording or any information storage and retrieval system now known or to be invented, without written permission from the publisher.

For photocopying of material in this volume, please pay a copying fee through the Copyright Clearance Center, Inc., 222 Rosewood Drive, Danvers, MA 01923, USA. In this case permission to photocopy is not required from the publisher.

ISBN 978-981-4411-11-0 (Hardcover)
ISBN 978-981-4411-12-7 (eBook)

Printed in the USA

Contents

Preface   xi

1. **Introduction**   1
   1.1   What is a Liquid Crystal?   1
   1.2   Brief History   4
   1.3   LC Classifications   12
   1.4   LC Phases   16
   1.5   Characterization Methods   21
        1.5.1   Differential Scanning Calorimetry (DSC)   22
        1.5.2   Polarizing Optical Microscopy   24
        1.5.3   X-Ray Diffraction   25
        1.5.4   Heating Glass Tubes   26
        1.5.5   Miscible with an LC Sample   27
        1.5.6   Other Methods   27

2. **Structures and Properties**   31
   2.1   Basic Structures   31
        2.1.1   Mesogenic Core Ring   31
        2.1.2   Central Linking Groups   33
        2.1.3   End Groups   34
        2.1.4   Lateral Substituents   35
   2.2   Some Physical Parameters   36
        2.2.1   Order Parameter   36
        2.2.2   Mechanical Properties (Elastic Constant)   36
        2.2.3   Electrical Properties (Dielectric Anisotropy)   38
        2.2.4   Magnetic Anisotropy (Diamagnetism)   38
        2.2.5   Optical Anisotropy (Birefringence)   39
        2.2.6   Fréedericksz Transition   40
   2.3   LC Alignment   41
        2.3.1   Rubbing Technique   42
        2.3.2   Microgroove Method   44
        2.3.3   Electric and Magnetic Fields   45
        2.3.4   Surfactant Dipping and LB Membrane   45
        2.3.5   Supramolecular Self-Assembly   47

		2.3.6	Oblique Evaporation	47
		2.3.7	Ionic and Plasma Beams	50
	2.4	Photoalignment		50
		2.4.1	Photoisomerization (Command Surface)	51
		2.4.2	Photocrosslinking (or Linearly Photodimerized) Reaction	53
		2.4.3	Photodegradation (Photoactive Polyimide)	56
3.	**Light and Liquid Crystals**			**63**
	3.1	Photochemical Reactions		64
	3.2	Photoinduced Alignment and Reorientation of LCs		68
		3.2.1	Photoresponse to Linearly Polarized Light	68
		3.2.2	Thermal Effect on Photoalignment	71
		3.2.3	Thermal Enhancement of Photoalignment in Photocrosslinkable LCs	75
		3.2.4	Photoresponse to Unpolarized Light	79
	3.3	Photomodulation of LCs		83
4.	**Low-Molecular-Weight Liquid Crystals**			**91**
	4.1	Photoinduced Phase Transition		91
		4.1.1	Photoinduced Phase Transition in Pure Photochromic LMWLCs	92
		4.1.2	Photoinduced Phase Transition in Azobenzene-Doped LCs	93
		4.1.3	Photoinduced Phase Transition in LCs Doped with other Dyes	98
		4.1.4	Photoinduced Phase Transition in LMWLCs–Polymer Composites	100
	4.2	Phototunning of Cholesteric LCs		103
		4.2.1	General Principles	103
		4.2.2	Phototuning CLCs with Photoisomerization	104
		4.2.3	Phototuning with other Chromophores	106
		4.2.4	Possible Applications	107
	4.3	Photochemical Flip of Polarization of Ferroelectric LCs		110

4.4	Phototriggered Sol–Gel Transition in LMWLC Organogels		114
4.5	Photocontrolled Orientation by Photophysical Processes		118
4.6	Photodriven Motion of LMWLCs		125

5. Liquid Crystal Polymers — **133**

5.1	Photochemical Phase Transition		133
	5.1.1	Copolymers and Polymer Composites	133
	5.1.2	Homopolymers	136
5.2	Photoinduced Cooperative Motion		143
5.3	Photoinduced Large Change in Birefringence		145
5.4	Polarized Electroluminescence (EL)		149
5.5	Holographic Applications		151
	5.5.1	Holographic Recording	151
	5.5.2	Recording Gratings with LC Alignment Changes	152
	5.5.3	Recording Gratings with Photoinduced Phase Transition	155
	5.5.4	Subwavelength Gratings	156
	5.5.5	Mechanically Tunable Gratings	157
	5.5.6	Bragg-Type Gratings	158
5.6	Other Applications		160
	5.6.1	Photocontrol of Functional Materials	160
	5.6.2	Photocontrolled Nanostructures	161
	5.6.3	Photorewritable Paper	161
	5.6.4	Photoswitching of Gas Permeation	163
	5.6.5	Photodriven Motions	163

6. Liquid Crystal Elastomers — **175**

6.1	Preparation of LCEs		176
	6.1.1	Two-Step Method	177
	6.1.2	One-Step Method of Direct Crosslinking of Linear LCPs	177
	6.1.3	Polymerization of LC Mixture of Monomers and Crosslinkers	178
	6.1.4	Physical Crosslinking Method	179
6.2	Photochemical Phase Transition		180
6.3	Photoinduced Contraction and Expansion		181
6.3	Photoinduced 3D Motions		184

		6.3.1	Photoinduced Bending	184
		6.3.2	Precise Control of Photoinduced Bending	187
		6.3.3	Effect of Order of Mesogen on Photoinduced Bending	190
		6.3.4	Effect of Light Polarization on Photoinduced Bending	192
		6.3.5	Effect of LCE Structures on Photoinduced Bending	192
	6.4	Microscale LCE Actuator		194
		6.4.1	LCE Fibers	194
		6.4.2	LCEs for Artificial Cilia	196
		6.4.3	LCEs with Micromolding	196
		6.4.4	Photocontrol of Surface-Relief Formation	197
	6.5	Novel LCE Materials and Photomechanical Ways		198
		6.5.1	Recyclable Hydrogen-Bonded LCEs	198
		6.5.2	Dye-Doped LCEs	199
		6.5.3	Twisted LCEs	200
		6.5.4	Hummingbird Movement	201
	6.6	LCE-Laminated Films		201
7.	**Liquid-Crystalline Block Copolymers**			**211**
	7.1	Synthesis of Well-Defined LCBCs		212
		7.1.1	Direct Polymerization	213
		7.1.2	Post-Functionalization	216
		7.1.3	Supramolecular Self-Assembly	217
		7.1.4	Special Reactions	217
	7.2	Phase Diagram of LCBCs		218
	7.3	Structures and Properties of LCBCs		222
		7.3.1	Effect of Microphase Separation on LC Phases	223
		7.3.2	Effect of Non-LC Blocks	225
		7.3.3	Effect of LC Blocks	228
	7.4	Control of Microphase Separation		230
		7.4.1	Thermal Annealing	232
		7.4.2	Mechanical Rubbing	234
		7.4.3	Photoalignment	236
		7.4.4	Electric and Magnetic Fields	240

		7.4.5	Other Methods	243
	7.5	Applications		245
		7.5.1	Enhancement of Surface-Relief Gratings	245
		7.5.2	Enhancement of Refractive-Index Modulation	246
		7.5.3	Nanotemplates	247
		7.5.4	Microporous Structures	251
	7.6	Outlooks		252
Index				259

Preface

It is well known that liquid crystal (LC) is a state between an ordered crystal and an isotropic liquid. Since its discovery in 1888, it has greatly influenced the daily life, and its use does not limit to flat-panel displays such as televisions and personal computers. The LC features endow materials in an LC state with interesting functionalities such as self-assembly, controllable fluidity with a long-range order, molecular and supramolecular cooperative motion, large birefringence and anisotropy in various physical properties (optical, mechanical, electrical, and magnetic), alignment change induced by external fields at surfaces and interfaces, and macroscopic deformation in response to stimuli such as electric, magnetic, photo, thermal, and mechanical forces.

Taking advantage of the recent progress in materials chemistry and physics, we can conveniently design and prepare photoresponsive LCs by the incorporation of photochromic molecules with LC materials. For instance, photoisomerizable azobenzene and photocrosslinkable cinnamate are typical photochromic groups for designing various light-active LCs. By controlling the wavelength, intensity, polarization, phase retardation, and interference of actinic light, one can photomodulate the LC unique characteristics such as photoinduced phase transition, photocontrolled alignment, and phototriggered molecular cooperative motion, giving rise to their photonic applications. Among the green and neat energy sources, light is particularly fascinating because it provides precise and reversible control in a noncontact way. LC materials with photoactive functions can be easily photo-manipulated into hierarchical structures with contrivable patterns, which offers a great opportunity for advanced functional materials beyond display applications. Combining these photoresponsive properties with three-dimensionally crosslinked elastomers may provide LC actuators with photomechanical and photomobile properties, directly converting the actinic light energy into mechanical work. Besides, the incorporation of supramolecular cooperative motion of LCs with microphase separation of block copolymers enables the well-defined LC block copolymers to exhibit controllable regular nanostructures in a macroscopic scale with excellent reproducibility and mass production. These newly developing aspects of photocontrollable LC actuators have opened up their applications ranging from displays and photonics to photodriven devices as well as nanotechnology.

This book comprises seven chapters. Chapters 1–3 present the basic knowledge of LCs and photoresponsive LCs, which new researchers will find useful for getting familiar with this field. Chapters 4–7 introduce photoresponsive LCs in the states of low-molecular-weight materials or small molecules, polymers or macromolecules, elastomers, and block copolymers, respectively. These advanced chapters may serve as a reference for experienced researchers to keep up with the current research trends. This book can be used as a textbook or a general reference for undergraduate and graduate students as well as primary researchers in both academia and industry in related fields such as organic chemistry, materials science and engineering, macromolecular engineering, supramolecular science and technique, electronic engineering, photonics, and nanotechnology. Hopefully, this book will help readers acquire professionally valuable knowledge and follow closely the pace of the newly developing area in photoresponsive LC materials.

Finally, I would like to express my gratitude to Stanford Chong of Pan Stanford Publishing for inviting me to write this book. I would also like to thank graduate students of Nagaoka University of Technology, Japan, and Peking University, China, who gave me valuable suggestions on my lectures on advanced LC materials, which form the basis of this book. I am indebted to my wife, Wei, and my son, Haoyang, for their great support, understanding, and encouragement, without which I would not have been able to write this book.

Haifeng Yu
Winter 2014

Chapter 1

Introduction

The history of human society is always accompanied by the development of novel materials, from the period of natural stone, the era of bronze, and the epoch of iron, to the modern times of synthetic materials and promising advanced functional materials. Liquid crystals (LCs) are one of such kinds of materials that greatly influence our daily life of mankind, which are not limited to displays such as televisions (TVs) and personal computers (PCs).

1.1 What is a Liquid Crystal?

Everyone knows LC displays (LCDs), which dominate the present market of flat panel displays in the world. But LCs should not be confused with LCDs. Generally, LCs are the so-called fourth state of materials, following the three states of materials: solid, liquid, and gas. In fact, many other states of materials like plasma and supercritical fluid have been discovered, as shown in Fig. 1.1.

Being an intermediate phase of matter, LCs show characteristics of both controllable mobility of an isotropic liquid and ordered regularity of a crystalline solid (Fig. 1.2). On the one hand, LCs can flow like water, which ensures their free movement and quick response to external stimuli. LC phases are the counterintuitive combination of fluid-like molecular mobility and crystal-like long-range ordering. Therefore, LCs can be regarded as an ordered liquid

---

Dancing with Light: Advances in Photofunctional Liquid-Crystalline Materials
Haifeng Yu
Copyright © 2015 Pan Stanford Publishing Pte. Ltd.
ISBN 978-981-4411-11-0 (Hardcover), 978-981-4411-12-7 (eBook)
www.panstanford.com

or a fluid crystal. Orientational order always exists and translational order often appears as well, the latter being of one-dimensional (1D), 2D or 3D type (in the 3D case, the order is, however, only short-range in at least one of the 3Ds). Among the translationally ordered LCs, a rich subdivision of generic classes into different phases has been carried out with subtle differences in terms of molecular organization, which leads to a broad spectrum of symmetries with important consequences for their physical properties.

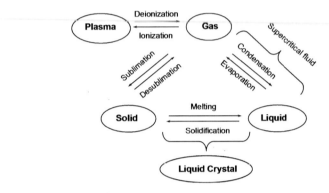

Figure 1.1 Summary of states of materials.

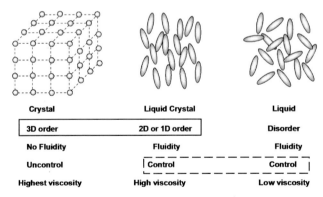

Figure 1.2 Simple description of the three states of crystal, LC, and liquid states.

Among millions of LC materials, 4-n-pentyl-4′-cyanobiphenyl (5CB) is one of the most popular, and its structure and properties are summarized in Fig. 1.3. From chemists' view, 5CB is a biphenyl moiety with substituent of a cyano moiety and one n-phenyl group in both ends. But it is a rod-like ellipsoid from viewpoints of physicists and engineers. 5CB crystals melt at 24°C from their crystalline state. The melted 5CB is a turbid liquid with fluidity like water, which is different from other materials exhibiting a transparent liquid at a temperature higher than their melting points. A beautiful Schlieren texture with strong birefringence can be observed with polarizing optical microscopy (POM) in this state. On heating the cloudy liquid further up to 35°C, 5CB converts to a transparent liquid as shown in Fig. 1.3. Under this circumstance, the birefringence disappears and only a dark image is obtained upon POM measurement. This "melting point" is the LC-to-isotropic phase transition temperature, which is also called "clearing point".

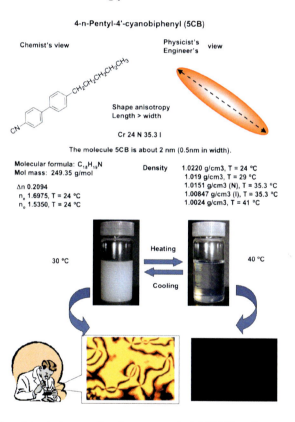

Figure 1.3   A typical example of one nematic LC (5CB) and its properties (Cr, crystal; N, nematic LC phase; I, isotropic phase).

1.2 Brief History

The understanding of the fascinating LC phenomena is a gradually advancing process. Even before the discovery of LCs, LC materials are being used since the discovery of soap. In the early 1850s, Rudolf Virchow described a soft, floating substance from nerve core, and named it "myelin" (Fig. 1.4). In 1850, a German biochemist, M. Heintz reported that stearin melted from solid state to a cloudy liquid at 52°C, which changed to opaque at 58°C and to a clear liquid at 62.5°C. In 1857, Mettenheimer discovered that myelin was birefringent since it is a lyotropic LC known today. At this stage, scientists in many fields of chemistry, physics, biology, and medicine observed that several materials behaved strangely at temperatures near their melting points. It was demonstrated that the optical properties of these materials changed discontinuously at elevated temperatures. However, the word "liquid crystal" did not appear until 1888.

Rudolf Ludwig Karl Virchow
(1821-1902)

German doctor, anthropologist, pathologist, prehistorian, biologist and politician.

C. Chr. Fr. Von Mettenheimer
(1824-1898)

Physician, opthamologist, microscopist Knighted for his work in anatomy, Physician, pathologist,scientist, clinical medicine and public welfare.

Figure 1.4 Two scientists who contributed to the gradual understanding of LC properties before the discovery of LCs in 1888.

It is well known that the discovery of LCs is attributed to the great work of Reinitzer and Lehmann [Reinitzer, 1888; Lehmann, 1890]. In 1888, the Austrian chemist, Friedrich Reinitzer, working in the Institute of Plant Physiology, University of Prague, discovered a strange phenomenon. When he heated cholesteryl benzoate (Fig. 1.5) to study its correct formula and molecular weight, he found that the compound melted at 145.5°C to form a turbid fluid and then appeared to melt again into a transparent phase at about 178.5°C. He also observed birefringence and iridescent colors when the temperature was set between these two "melting points". However, Reinitzer did not concretely know this discovery. Then he wrote a letter to Otto Lehmann, a Germany physicist, for further discussion since Austria allied with German at that time. Lehmann is the first to use a POM equipped with a hot stage, which later became one of the standard instruments for LC researches (Fig. 1.5). He confirmed the existence of "the lowing, soft crystalline state" in Reinitzer's samples [Lehmann, 1890]. It is the pioneer work of the two grandfathers of LC science that opened the door to fundamentally understand the nature of this new, beautiful, and mysterious phase of matter, with fascinating and unique optical properties. Since then, the magical world of LC is open to humankind.

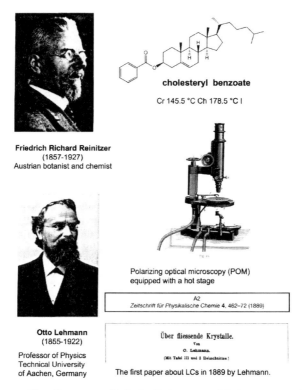

Figure 1.5 The discovery of LCs by Reinitzer and Lehmann in 1888 (Cr, crystal; Ch, cholesteric; I, isotropic phase).

This new discovery was greatly challenged by the scientific community, and some famous scientists claimed that the newly discovered state probably was just a mixture of solid and liquid components. But in the years between 1910 and 1930 conclusive experiments and early theories supported the LC concept, and at the same time new types of material states of order were discovered. The first synthesized LC compound of azoxybenzene derivatives was carried out by Gatterman in his laboratory (Fig. 1.6). Then Daniel Vorläder, another Germany scientist dominated the LC research scene for three decades beginning in 1905 [Vorländer, 1905]. He synthesized most of the LC compounds known until his retirement in 1935. He is the first to study the influence of molecular structure on the performance of LC phases. In 1922, Georges Friedel suggested a classification method to name different LC phases called nematic, smectic, and cholesteric with the Greek terms [Friedel, 1922], which are still used today.

Then a great scientist, C. W. Oseen, who nominated Albert Einstein for the Nobel Prize in 1921, formulated the fundamentals of the elasticity theory of LCs (Oseen elasticity theory). The theory was employed on the continuum theory by England's F. C. Frank. This theory became one of the fundamental theories in LCs today. In this formalism, an LC material is treated as a continuum, and molecular details are entirely ignored. The LC distortions are commonly described by the Frank free-energy density. One can identify three types of distortions that could occur in an oriented LC sample:

(1) Twists of the materials, where neighboring molecules are forced to be angled with respect to one another, rather than aligned;

(2) Splay of the material, where bending occurs perpendicular to the director;

(3) Bend of the material, where the distortion is parallel to the director and molecular axis.

These distortions incur an energy penalty. They are distortions that are induced by the boundary conditions at domain walls or the enclosing container. The response of the materials can then be decomposed into terms based on the elastic constants corresponding to the three types of distortions. Elastic continuum theory is a particularly powerful tool for modeling LC devices. In the 1930s, the Fréedericksz transition was found as a phase transition in LCs produced when a sufficiently strong electric or magnetic field is applied to an LC in a homogeneous state [Fréedericksz & Repiewa, 1927].

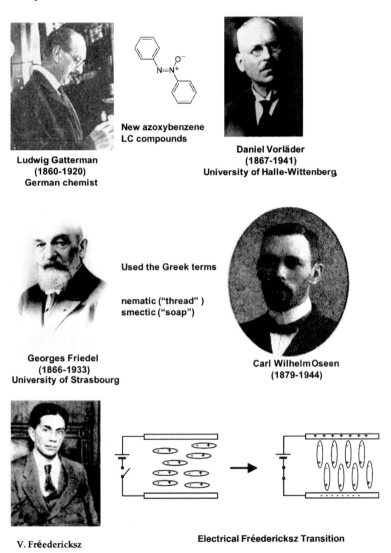

Figure 1.6 Famous people in the development of LC science since 1888.

Maier–Saupe mean field theory

Alfred Saupe
(1925-2008)

CHEMICAL REVIEWS

Subscriber access provided by

The Mesomorphic State - Liquid Crystals
Glenn H. Brown, and Wilfred G. Shaw
Chem. Rev., 1957, 57 (6), 1049-1157 • DOI: 10.1021/cr50018a002

Glenn Brown
(1915-1995)

Figure 1.7 Two famous scientists who contributed greatly to the LC science.

Then, two world wars and their aftermaths was a hard period for modern LC researches. General interest in LCs was at a fairly low level and only relatively few scientists devoted to extending the knowledge of LCs. After the world wars, markedly rapid development was achieved in this field, and some originally technological applications were invented based on LC properties. In 1958, Alfred Saupe studied the nematic LC-to-isotropic phase transition with his supervisor Wilhelm Maier, leading to the Maier–Saupe theory, which is another well-known basic LC theory. It includes contributions from an attractive intermolecular potential from an induced dipole moment between adjacent LC molecules. The anisotropic attraction stabilizes parallel alignment of neighboring molecules, and the theory then considers a mean-field average of the interaction.

In 1957, Glenn Brown, an American chemist, published a review article in *Chemical Reviews* on the LC phase and subsequently, sparked an international resurgence in LC researches [Brown & Shaw, 1957]. The early electro-optical indications led to today's LCD industry. The requirement of new applications stimulated researches and a huge amount of fund had been spend in LC-involved areas. The blooming of LC researches resulted in the First International Liquid Crystal Conference (ILCC) held at Kent State University, Ohio, USA, in 1965, with an attendance of 90 delegates. Now, it has become one of the most important international conferences, held every two years.

The first application of cholesteric LCs (CLCs) to human thermography was developed as a cost-effective way for doctors to diagnose diseases beneath the skin (Fig. 1.8). The most important application of LCs in display fields appeared subsequently. In 1968, George Heilmeier, Radio Corporation of America (RCA), first developed LCD devices based on dynamic scattering, as shown in Fig. 1.9.

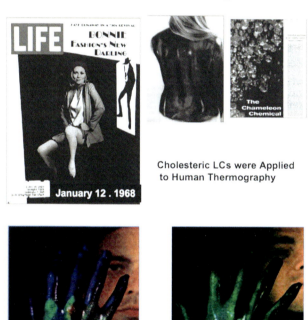

Figure 1.8 The first application of LCs based on cholesteric LCs.

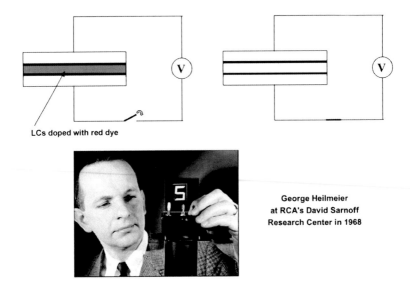

Figure 1.9 Scheme of LCD devices based on dynamic scattering.

Then, Helfrich and Schadt (Europe) and Fergason (USA) independently applied for patents on the concept of the twisted nematic (TN) LCD. LCD manufacturing for watch and calculator displays began in the United States (RCA and International Liquid Xtal Company (ILIXCO)). Steadily, the research for LC applications broadened its view and the milestone in this field might be the discovery of the 4-alkyl- and 4-alkoxy-4′-cyanobiphenyls (Fig. 1.10) by Gray and coworkers in 1973. High quality, reliable LCDs could be built with these LC materials. Japanese companies acquired the technology through licensing agreements, and they developed such techniques and aroused the modern LCD industries.

Thanks to Reinitzer, Lehmann, and their followers, we presently know that literally thousands of substances have a diversity of other states. Some of them have been found very applicable in several technical innovations. In 1977, a new type of discotic LC (hexahydroxbenzene-n-alkanoates) was first synthesized by Chandresekhar's research group [Chandresekhar et al., 1977]. In 1960s, a French theoretical physicist, Pierre-Gilles de Gennes, who had been working with magnetism and superconductivity, turned his interest to LCs and soon found fascinating analogies between LCs and superconductors as well as magnetic materials. The Noble Prize in physics 1991 was bestowed to him for discovering "the methods developed for studying order phenomena in simple systems can be generalized to more complex forms of matter, in particular to LCs and polymer materials" [Gennes, 1995]. This is a milestone of LC development, and LCs become an interdisciplinary research area for chemists, physicists, biologists, engineers, and other fields.

Figure 1.10 Three typical scientists who greatly influenced LC researches and applications.

Recently, photochromic molecules such as photoisomerizable azobenzene, and photocrosslinkable cinnamate have been integrated into LC materials, which enables one to precisely light-manipulate the unique properties of the designed materials by photoinduced phase transition, photocontrolled alignment, phototriggered molecular motion, and supramolecular cooperative motion (SMCM). These give birth to their photonic applications. Connecting with 3D-crosslinked elastomers, photoresponsive LC materials show photomechanical and photomobile properties, directly converting light energy into mechanical one. Combining SMCM with microphase separation of well-defined LC block copolymers (LCBCs), they exhibit light-controllable ordered nanostructures in a macroscopic scale, with excellent reproducibility and mass production. The advances in photoresponsive LC materials will be discussed in the following chapters.

1.3 LC Classifications

The various LC phases (also called mesophases) can be characterized by the kind of molecular ordering. One can distinguish positional order and orientational order, and moreover, order can be either short-range or long-range. Most of thermotropic LCs have an isotropic phase at a high temperature because heating will eventually drive them into a conventional liquid phase with fluid-like flow behavior. Under other conditions (for instance, lower temperature), an LC might inhabit one or more phases with significant anisotropic orientational structures and short-range orientational order while still having flow properties.

As previously described, cholesteryl benzoate studied by Reinitzer and Lehmann are directly purified from nature [Reinitzer, 1888; Lehmann, 1890]. But most of LC materials are man-made or synthesized by chemists in laboratory. To show LC properties, materials generally have anisotropic molecular structures. This concept has been widely accepted by most of researchers in this field. Based on different aspects, like molecular shape, molecular weight, and LC formation conditions, LCs can be accordingly classified into a lot of types. For instance, considering geometrical structures, LCs are generally rod-like, disc-like, bowl-like, and banana (bent-core) shaped molecules, as shown in Fig. 1.11.

Molecules, which form an LC phase in bulk by packing in an ordered fashion, may change their packing arrangement as a function of concentration in a solvent (lyotropic LCs) or as a function of temperature (thermotropic LCs). Therefore, LCs can be divided into three types, thermotropic, lyotropic, and amphotropic, according to their formation conditions [Tschierske, 2002]. As shown in Fig. 1.12, substances that exhibit LC states at a certain temperature range are named thermotropic LCs. Usually, there are two changes when the thermotropic LCs are heated up. The first change happens at its melting point with the phase transition from a crystalline solid to an LC phase. When the temperature is further increased, the phase of the substance changes from an LC phase to an isotropic liquid phase. The point at this temperature is also called the "clearing point" (Fig. 1.12). An LC phase may be formed either by cooling an isotropic liquid below the clearing point or by heating a solid crystal above its melting point. Some LC materials have a rich phase diagram, exhibiting several different smectic phases at different temperatures. Since most of the photoresponsive LCs are thermotropic, only thermotropic LCs will be introduced in detail in this book.

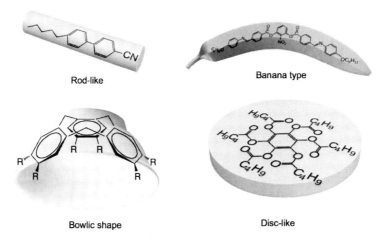

Figure 1.11 Examples of various geometrical structures of LC molecules.

Compounds that exhibit an LC phase by addition of an appropriate solvent (the most conventional one is water) in suitable amounts are called lyotropic LCs, which often forms over a range of concentration. In formation of the lyotropic LC phases, solvent molecules fill the space around the compounds to provide fluidity to the system. In contrast to thermotropic LCs, these lyotropic LCs have another degree of freedom of concentration that enables them to induce a variety of different phases. The fluid anisotropy often results from polar head-group packing of amphiphilic molecules. In addition, the influence of temperature gives additional degree of freedom, but is of secondary importance. The classical representatives of this type of LCs are hydrocarbon foams and soap (surfactant) [Laughlin,

1994]. The mesophase morphologies for lyotropic surfactant LCs are summarized in Fig. 1.13.

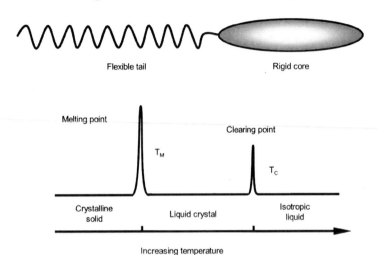

Figure 1.12 Thermotropic LCs and schematic illustration of their thermal analysis.

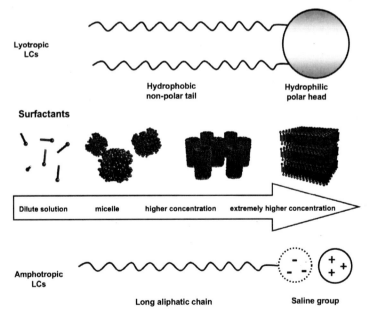

Figure 1.13 Scheme of lyotropic and amphotropic LCs.

In some cases, molecules can show both thermotropic and lyotropic LC phases, which is described as amphotropic behavior. The amphotropic compounds can exhibit an LC phase through either heating above the melting point or the addition of solvents. However, the mesophases observed in these two regimes are often quite different owing to the temperature, shapes, and chemical structures [Borisch et al., 1998]. Examples of such amphotropic molecules are amphiphilic polyhydroxy compounds, carbohydrate derivatives, and block copolymers.

As far as molecular weights are concerned, LCs are often low-molecular weight (LMW) or small molecules, oligomers, and high-molecular weight polymers or macromolecules. Due to low viscosity and quick response, LMW LCs or their mixtures are generally used in LCDs. According to the positions of mesogens in polymers, LC polymers (LCPs) can be classified as main chain and side chain LCPs, as shown in Fig. 1.14. When mesogenic units are linked in main chain, the polymers are called main-chain LCPs. Due to the spacer between the mesogenic unit and the main chain, LCPs were formed through two methods. Mesogens in the first type of main-chain LCPs are connected by stiff spacers, and the second type of main-chain LCPs contains flexible bridges. The participation of flexible spacers dramatically changes the properties of LCPs. The theoretical basis for these LCPs has been studied widely [Matheson & Flory, 1981; Wang & Zhou, 2004].

If a mesogen is introduced into the polymer side chain, the polymers were named side-chain LCPs. When a spacer exits between the polymer backbone and the side mesogen, they are classified into two main categories (Fig. 1.14): the end-on LCPs, in which the side chains are terminally attached to backbone, and the side-on LCPs. If no spacer or only a very short spacer exists, the mesogenic units would form a dense "jacket" around each chain backbone because of their high population in and around the backbone. As a result, they are both bulky and rigid. This kind of LCP was called mesogen-jacketed [Zhou et al., 1987]. Since such kinds of LCPs with mesogenic units attached laterally to main chains, their properties are more like those of main chain LCPs.

Considering the architecture of polymer chains, LCPs also can be homopolymers, block copolymers, graft polymers, crosslinked polymers, cyclic, dendrimers, and hyperbranched-LCPs (Fig. 1.15).

This macromolecular architecture often shows strong effect on the LC properties of LCPs.

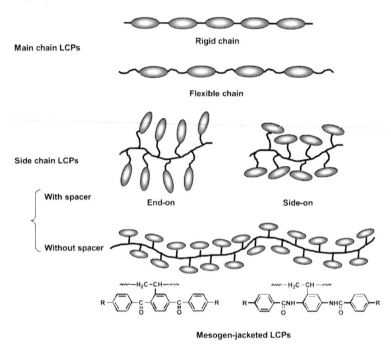

Figure 1.14 Schemes of main-chain and side-chain type LCPs.

1.4 LC Phases

According to orientational and positional orders, thermotropic LCs are classified as three main classes, nematic, cholesteric (or chiral nematic), and smectic (Fig. 1.16). In general, nematic molecules are centrosymmetric, and the nematic phase has only an orientational order (no positional order). Thus, the molecules are free to flow and their center of mass positions are randomly distributed as in a liquid, but still maintain their long-range directional order. The cholesteric phase not only has an orientational order similar to the nematic phase in all physical properties, but also possesses an arrangement in a helical manner. This spontaneous helical director configuration makes it more difficult to reorient a CLC than a nematic one. The smectic phases possess a positional order, that is, the position of the molecules is correlated in some ordered pattern. Smectic LCs form well-defined layers that can slide over one another in a manner similar to that of soap. Several forms of smectic phases have been discovered and labeled from smectic A to smectic K currently. They are separated by their tilt angles (with respect to the plane normal) and packing formation. Some typical textures of these LCs are also shown in Fig. 1.16.

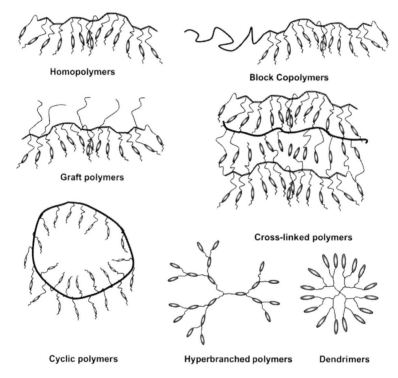

Figure 1.15 LCPs with diverse macromolecular architectures.

Introduction of a chiral center into an LC molecule may bring about interesting properties. As shown in Fig. 1.17, 5CB is a typical nematic LC (Fig. 1.3), but a chiral nematic (or cholesteric) LC phase is obtained when a chiral molecule is present. Similarly, chiral smectic A (or ferroelectric) LCs can be achieved by this way. Another method to endow LCs with chirality is doping or mixing chiral molecules into non-chiral LCs. In some cases, non-chiral molecules are capable of forming chiral LC phases. For example, molecules with a bent core (or banana-shaped molecules) can build polar, and even chiral LC structures, showing a variety of new LC phases (B1–B7).

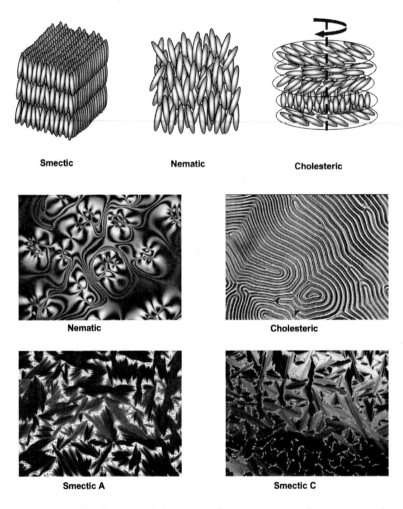

Figure 1.16 LC phases and their typical textures upon observation with POM.

Blue phases are one of the most interesting self-organization structures in the field of LC researches. Figure 1.18 shows typical examples of blue phase, which generally appears between a CLC phase and an isotropic liquid. Blue phase LCs usually possess a very narrow temperature range of only a few degrees of Kelvin. In this phase,

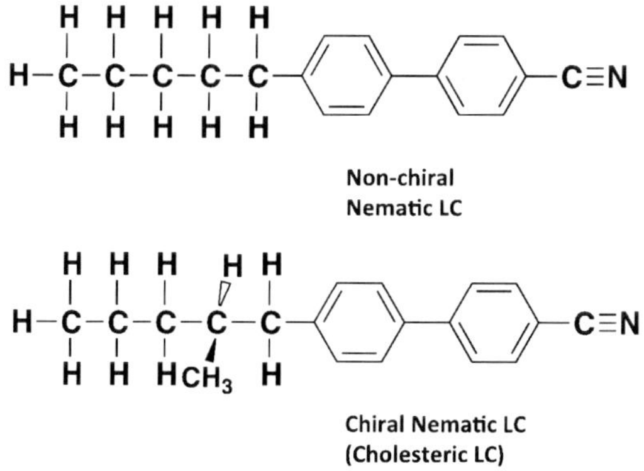

LC Phases | 19

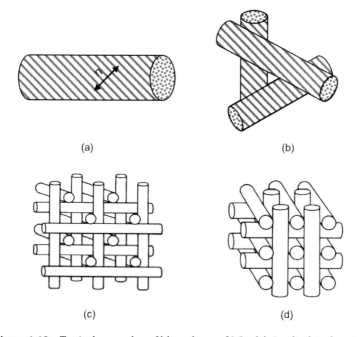

Figure 1.17 One example of chiral LC phase obtained by introducing a chiral molecule.

Figure 1.18 Typical examples of blue phase of LCs. (a) A cylindrical region of double-twist structure; (b) Intersection of three double-twist cylinders which form a defect; (c) Arrangement of double-twist cylinders which may exist in BPI; and (d) Arrangement of double-twist cylinders which may exist in BPII.

the molecular directors rotate in a helical fashion perpendicular to lines that form a number of helical axes in different directions. It is sometimes called a double-twist structure even though an unlimited number of axes are present, which is more stable than the single-twist structure found in CLCs. The double-twist structures are limited in all directions to the distance from the centerline where the twist amounts to 45° and a double-twist cylinder causes. Such a cylinder is more stable because of its small radius compared to the same volume filled with only one single-twist of CLCs. Blue phases have a regular 3D cubic structure of defects. If the spacing between the defects is in the range of the wavelength of light, interference occurs in a certain wavelength range for light reflected from the lattice (especially blue-colored light). These double-twist cylinders form a cubic lattice wherein the defects occur at the contact point of the cylinders.

Generally, LCs are typically rod-like anisotropic materials with parallel orientation and limited positional order, forming one or several mesophases between the solid crystal and isotropic liquid phase. Both the refractive index and the static dielectric constant depend on the orientation of the director and the electric field, which causes the birefringence and dielectric anisotropy, respectively. This phenomenon is exceptional for blue phase LC materials, which appears in chiral systems in the temperature range just below the isotropic liquid phase. These phases are optically isotropic and they are not birefringent. However, they may show colors due to the selective reflection of circularly polarized light. Till now, three thermodynamically stable blue phases have been observed on cooling from the isotropic phase to the chiral nematic phase. They are the colorful BPI, BPII, and the misty blue BPIII, which are separated by the first-order transitions. Both BPI and BPII have long-range orientational order with 3D cubic symmetry. BP1 is body-centered cubic and BPII is simple cubic, while BPIII is only isotropic.

Recently, it has been demonstrated that some new materials possess blue phases with enhanced thermal stability ranging from 1°C to 3°C and the flexoelectricity effects observed in them are profound [Kikuchi et al., 2002]. This report is of interest, as any practical application of some electro-optical phenomenon in a blue phase will require an operating temperature regime well in excess of the few degree range characteristic of conventional materials.

Coles reported that the stabilization of these blue phases over a wide temperature range of 60°C in a mixture of three homologous dimers mixed in equal proportion with a small amount of chiral dopant [Coles & Pivnenko, 2005]. In May 2008, it was announced that Samsung had developed the first blue phase LCD panel.

The above-mentioned phase classification method belongs to LCs with rod-like molecules. However, some LCs are formed by disc-like molecules, which is called discotic LCs. Generally, there are three kinds of discotic LCs: nematic discotic, cholesteric, and columnar discotic phases (Fig. 1.19). Disk-shaped LC molecules can orient themselves in a layer-like fashion known as the discotic nematic phase. If the disks pack into stacks, the phase is called a discotic columnar. The columns themselves may be organized into rectangular or hexagonal arrays.

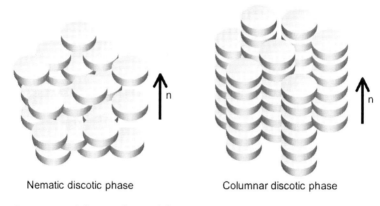

Nematic discotic phase   Columnar discotic phase

Figure 1.19 Scheme of typical discotic LC phases.

1.5 Characterization Methods

A complete characterization of LC materials should include at least two aspects: molecular structures and LC properties. Since the first aspect characterization is nothing more than non-LC materials, only methods mostly used for the characterization of LC phases will be briefly discussed here. The LC properties are most commonly analyzed using a combination of differential scanning calorimetry (DSC), POM, X-ray diffraction (XRD), miscibility investigation, and other special methods like Fourier transform infrared spectroscopy

(FTIR), nuclear magnetic resonance spectroscopy (NMR), small-angle neutron scattering (SANS), and so on. Thanks to the development of modern scientific researches, these complementary techniques together provide considerable information regarding the LC phases.

1.5.1 Differential Scanning Calorimetry (DSC)

Being one of the most important methods of thermal analysis, DSC is often applied in measurement of temperature and enthalpy in the phase transitions of LC materials. As shown in Fig. 1.20, two pans are simultaneously heated in two separated holders in this method. The first pan contains a certain amount of LC sample materials to be analyzed, while the second (empty) pan is used as a reference. The temperature of the two pans is changed at a constant rate and the difference in heat flow required to maintain both at the same temperature is measured.

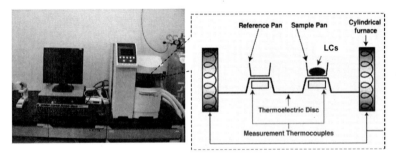

Figure 1.20 One picture of DSC instrument and its schematic illustration of the structure of the measurement cell.

At a phase transition, excess heat is absorbed (on heating) or released (on cooling) by the sample, which results in the appearance of a peak on a plot of temperature versus heat flow. These plots are referred to as endotherms when the sample is heated and exotherms when it is cooled down, as shown in Fig. 1.21. These plots produce two very useful pieces of information regarding phase information of LC materials: the temperature at which phase transitions occur and the enthalpy of those transitions. The temperature at which a transition occurs is useful in predicting how stable a state will be. A higher transition equates to a phase being more thermally stable.

The enthalpy is simply the area under the curve multiplied by a calorimetric constant that depends on the instrument used. DSC is very sensitive and so can detect the small changes in heat that transitions from one LC phase to another produce. The magnitude of the enthalpy of transition reveals how large the change in molecular order is between the two phases. The crystal-to-LC transition tends to have large enthalpies of transitions (about 20–100 kJ/mol), whereas the enthalpy of LC-to-isotropic (at the clearing point) or LC-to-LC phase transitions tend to be smaller (about 1–10 kJ/mol).

There are two types of thermotropic LCs results from DSC measurement. As shown in Fig. 1.21, the first is called enantiotropic LCs where the LC phase can be obtained from either lowering the temperature of a liquid or raising the temperature of a solid. The second is monotropic LCs, which is an irreversible process where the LC phase can only be reached from one direction (upon cooling) in the thermal cycle.

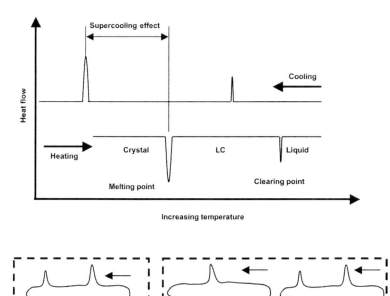

Figure 1.21 One typical DSC curve (above) of LC materials and two kinds of thermotropic LCs results from DSC measurement (below).

In a DSC trace of a crystalline or LC materials, there is normally a temperature displacement in the formation of a crystalline phase or a mesophase in the cooling cycle as compared with a heating cycle. This displacement is also called the supercooling effect. Compared with a crystalline material, the supercooling of an LC–LC or an LC–isotropic transition is typically small. This is a fast and facile way to judge the likely presence of an LC material by DSC.

1.5.2 Polarizing Optical Microscopy

Although POM observation is one of the most simple and convenient methods to characterized one LC sample, it can provide a great deal of information not available with other techniques. Therefore, it is one of the most popular methods in LC researches. Isotropic materials (e.g., water) often have the same refractive index in all directions. In contrast, many anisotropic materials have optical properties that vary with the orientation of incident light with respect to the axes of LCs. At this case, light will propagate at different rates in different directions through LC materials, because light travels through different directions at different velocities.

In observation with POM, the LC sample is placed between a pair of two-crossed polarizers, that is, two polarizers that are rotated by 90° with respect to one another. Figure 1.22 shows schematic illustration of the principle of a POM. One polarizer is between the light source and the LC sample, while the other (also called analyzer) is between the sample and the observer. Changes in the image viewed relative to that of a regular microscope arises from the interaction of linearly polarized light (LPL) with a birefringent LC material. Optical textures arise because the LCs are not perfectly homogenous and defects or deformations occur as the LC patterns grow from an isotropic phase. Viewed under POM, different types of LCs show their characteristic textures, which are often used to identify LC phases. For instance, a nematic phase shows a thread-like Schlieren texture and smectic A shows fan-like texture, as shown in Fig. 1.16.

In addition, an LC phase is temperature dependent. The optical anisotropy disappears when the LC is at a temperature above its mesophase to isotropic transition temperature. Therefore, the clearing temperature can be roughly determined by following the appearance of a dark field under POM with a hot stage.

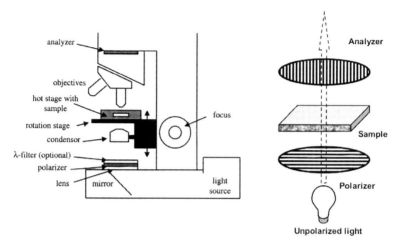

Figure 1.22 Structure and principle of POM measurement for LC observation.

1.5.3 X-Ray Diffraction

When fast electrons impinge on a matter, electrons are often ejected from the inner shell of atoms of target materials. The resultant vacancies will be filled by electrons falling from a higher energy shell. Since the energy levels of the different shells are quantized, a coherent ray is emitted in the form of an X-ray photon with energy equal to the difference in energy levels of the two shells. XRD is widely used to identify a smectic phase from a crystal, a nematic phase, and other smectic phases. X-rays are diffracted at crystal planes since the distance between the planes is comparable to the wavelength of the X-ray (Fig. 1.23).

Typical XRD data of LC samples are given in Fig. 1.23. In general, smectic LC samples often show sharp peaks at the small-angle area because of the 2D ordered structures. A nematic LC exhibits its wide peak at the wide-angle range due to its 1D ordering.

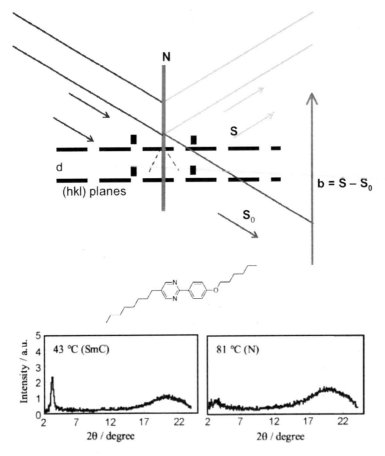

Figure 1.23 Mechanisms of XRD and one typical example of an LC compound. The incident (S_0 is the unit vector of the incident wave-normal) and diffracted ("reflected", S is the unit vector of the reflected wave-normal) X-ray beams on two parallels crystal lattice planes (hkl) separated by a distance dhkl. The vector $b = S-S_0$ is normal to the lattice plane (hkl) with $|b| = |S-S_0| = 2\sin\theta$.

1.5.4 Heating Glass Tubes

Heating a capillary tube containing one sample slowly might be one of the simplest methods to characterize one LC phase. Non-LC materials are melted into transparent liquids, whereas LC samples show non-transparent liquid because of the scattering of visible light. Due to their low viscosity, nematic LC materials flow down along the wall of glass tube. Smectic LC samples often stick to the glass wall because of their high viscosity. Bright colors often appear when heating CLC materials.

1.5.5 Miscible with an LC Sample

When you have one sample with a well-defined LC phase at hand, miscibility investigation might be the convenient method to check the LC phase of other unknown LC materials. The rule of miscibility for the identification of LC phases can be depicted in Fig. 1.24. If one unknown phase is continuously miscible with a reference phase over the entire concentration range, then they are equivalent. On the other hand, if they are not miscible, no conclusion can be drawn. For instance, if you can get miscible with mixing one nematic LC sample with uncertain materials, the materials should show a nematic phase. If phase separation occurs, the materials should not exhibit a nematic phase.

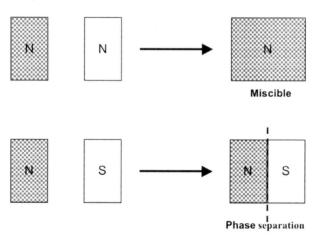

Figure 1.24 Measurement of LC phases by a miscible way with one known LC sample (N, nematic phase; S, smectic phase).

1.5.6 Other Methods

The usual way of characterizing LC phases is to identify the phase by POM, to establish the thermodynamic reversibility using DSC, and to elucidate the molecular arrangements using XRD. The other spectroscopic tools like FTIR, FT-Raman spectroscopy, small-angle neutron scattering (SANS), and refractive index measurements have also been adopted. FTIR can be used to bond conformation of LC materials. Polarized absorption spectra can supply order information of mesogenic orientation. The solution NMR spectroscopy is routinely used for structural characterization of mesogens in an isotropic phase, whereas the solid-state NMR spectroscopy has emerged as a powerful tool for molecular structural characterization of LCs in the mesophase [Domenici et al., 2007].

References

Borisch, K., Tschierske, C., Goring, P. and Diele, S. (1998). Molecular design of thermotropic liquid crystalline polyhydroxy amphiphiles exhibiting columnar and cubic mesophases of the normal type, *Chem. Commun.* pp. 2711–2712. DOI: 10.1039/A808271E.

Brown, G. H. and Shaw, W. G. (1957). The mesomorphic state-liquid crystals, *Chem. Rev.* **57**, pp. 1049–1157.

Chandrasekhar, S., Sadashiva, B. K. and Suresh, K. A. (1977). Liquid crystals of disc-like molecules, *Pramana J. Phys.* **9**, pp. 471–480.

Coles, H. J. and Pivnenko, M. N. (2005). Liquid crystal 'blue phases' with a wide temperature range, *Nature* **436**, pp. 997–1000.

Friedel, G. (1922). The mesomorphic states of matter, *Ann. Phys.* **18**, pp. 273–474.

Fréedericksz, V. and Repiewa, A. (1927). Theoretisches und experimentelles zur Frage nach der Natur der anisotropen Flüssigkeiten, *Zeitschrift für Physik Society* **42**, pp. 532–546.

Gray, G. W., Harrison, K. J. and Nash, J. A. (1973). New family of nematic liquid crystals for displays, *Electron. Lett.* **9**, pp. 130–131.

Gennes, P. G. and Prost J. (1995). *The physics of liquid crystals*, 2nd edition, USA: Oxford University Press.

Kikuchi, H., Yokota, M., Hisakado, Y., Yang, H. and Kajiyama, T. (2002). Polymer-stabilized liquid crystal blue phases, *Nat. Mater.* **1**, pp. 64–68.

Laughlin, R. G. (1994) *The aqueous phase behavior of surfactants*, London, UK: Academic Press.

Lehmann, O. (1890). Einige fälle von allotropie, *Z. Krist.* **18**, pp. 464–467.

Matheson, R. R., Jr. and Flory, P. J. (1981). Statistical thermodynamics of mixtures of semirigid macromolecules: Chains with rodlike sequences at fixed locations, *Macromolecules* **14**, pp. 954–960.

Reinitzer, F. (1888). Beiträge zur lenntnis des cholesterins, *Monatsh. Chem.* **9**, pp. 421–441.

Tschierske, C. (2002). Amphotropic liquid crystals, *Curr. Opin. Colloid Interface Sci.* **7**, pp. 355–370.

Domenici, V., Geppi, M. and Veracini, C. A. (2007). NMR in chiral and achiral smectic phases: Structure, orientational order and dynamics, *Prog. Nucl. Magn. Reson. Spectrosc.* **50**, pp. 1–50.

Vorländer, D. (1908). *Kristallinische-flüssige substanzen*, Stuttgart, Germany: Enke.

Wang, X. J. and Zhou, Q. F. (2004). *Liquid crystalline polymers*, World Scientific.

Zhou, Q. F., Li, H. M. and Feng, X. D. (1987). Synthesis of liquid-crystalline polyacrylates with laterally substituted mesogens, *Macromolecules* **20**, pp. 233–234.

Chapter 2

Structures and Properties

Generally, liquid crystals (LCs) are formed in rod-like organic molecules which interact through strongly anisotropic intermolecular forces. The LC properties are decided by their molecular structures. With the progress in LC researches, it has been becoming a common view that to show LC phases, the materials should meet the requirement of the following items:

(1) Anisotropic geometry, that means the structure should be either elongated (linear) in shape or inhomogeneous.
(2) Rigidity along the molecular axis.
(3) A big π-electron system and polarizablity of electron density.

2.1 Basic Structures

Typically, molecular structure of rod-like LCs are shown in Fig. 2.1. Four parts are often included, and they are the mesogenic core rings (B, B′), linking group (A), end groups (X, Y), and lateral substituents (Z, Z′).

2.1.1 Mesogenic Core Ring

Mesogenic core ring systems are one of the most important constituents of molecular structures in LC materials. Typical core rings are phenyl, biphenyl, terphenyl, and pheylcyclohexane groups

Dancing with Light: Advances in Photofunctional Liquid-Crystalline Materials
Haifeng Yu
Copyright © 2015 Pan Stanford Publishing Pte. Ltd.
ISBN 978-981-4411-11-0 (Hardcover), 978-981-4411-12-7 (Ebook)
www.panstanford.com

(Fig. 2.2). The kind of core ring groups strongly influences the LC properties, such as temperature, birefringence (Δn), dielectric constant ($\Delta \varepsilon$), etc.

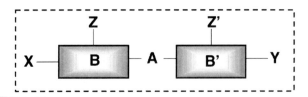

A – linking group
B, B' – ring systems, aromatic or saturated ring core
X, Y – end groups
Z, Z' – lateral substituents

Figure 2.1 General structures of rod-like LC molecules.

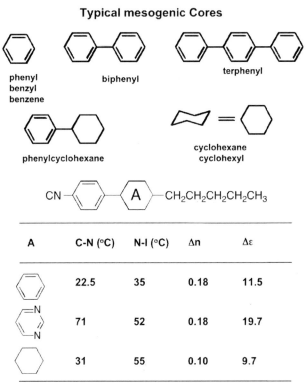

Figure 2.2 Influence of the mesogenic core ring on LC properties (C, crystal; N, nematic; I, isotropic phase).

2.1.2 Central Linking Groups

The central linking groups can lengthen the molecule and increase the length-to-breadth ratio. It greatly affects the linearity, planarity, and rigidity of LC molecules. Figure 2.3 shows common central linkages of mesogens. Most of the linkages are highly polarizable with the exception of steroid ring system and acid dimmer, but they are still rigid and linear. In general, the stronger the dipole or the easier the polarizability of the central group, the better is the thermal stability of the LC phase. It must be mentioned here that the azobenzene-containing compounds with photoisomerizable feature are often used for synthesizing photoresponsive LC materials.

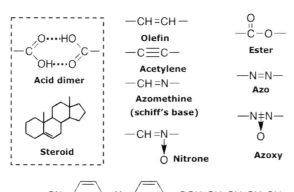

X	C–N(°C)	N–I(°C)
Azoxy	93	142.6
Azo	92	139
Acetylene	97	126
Olefin	87	96
Ester	78	94
Azomethine	62	93
Single bond	53	67.5

Figure 2.3 Influence of the central linking group on LC performances. (C, crystal; N, nematic; I, isotropic phase).

2.1.3 End Groups

The end groups usually play a role of extending the molecular axis (or long axis), which might improve the LC thermal stability. The presence of the dipolar terminal causes higher thermal stability of the mesogen, as shown in Fig. 2.4. In addition, nearly all the LC molecules contain at least one alkyl end chain. Short terminal chain favors nematic rather than smectic formation. A branched alkyl chain usually suppresses orthogonal smectic phases but can induce tilted smectic phases. The carbon number of alkyl end chain often shows so-called "odd–even effect" on the clearing point.

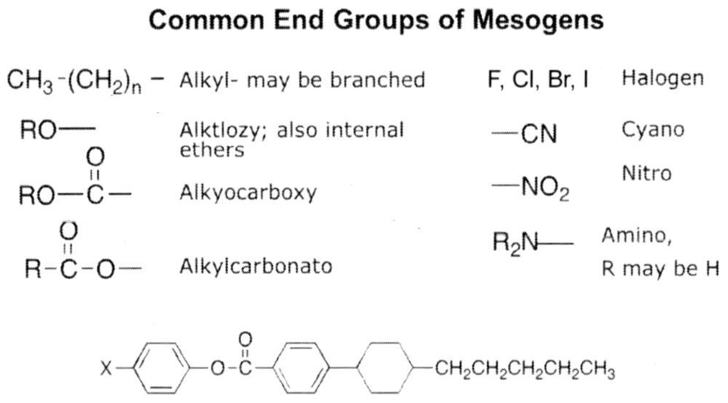

Common End Groups of Mesogens

$CH_3-(CH_2)_n-$	Alkyl- may be branched	F, Cl, Br, I	Halogen
RO—	Alktlozy; also internal ethers	—CN	Cyano
RO—C(=O)—	Alkyocarboxy	—NO_2	Nitro
R—C(=O)—O—	Alkylcarbonato	R_2N—	Amino, R may be H

X—⌬—O—C(=O)—⌬—⌬—$CH_2CH_2CH_2CH_2CH_3$

X	C–N (°C)	N–I (C°)
H	87.5	114.0
F	92.0	156.0
Br	115.0	193.0
CN	111.0	226.0
CH_3	106.0	176.0
C_6H_5	155.0	266.0

Figure 2.4 Influence of the end group on LC properties. (C, crystal; N, nematic; I, isotropic phase).

Even a small change in the end group may greatly change the LC properties. Figure 2.5 shows thermal properties of three azobenzene-containing monomers, which are often used to synthesize photoresponsive LC polymer materials. The acrylate monomer shows nematic LC phase upon cooling from its isotropic phase. However, no LC phase is observed for the methacrylate monomer although it has a similar structure to the acrylate monomer. For such molecular structures, acrylate monomer is easier to show an

LC phase than the methacrylate monomer. Possessing a symmetric structure, but the crosslinker does not exhibit any LC phases. Both of the polymers of the two monomers demonstrate nematic LC phase, which will be discussed in Chapter 5.

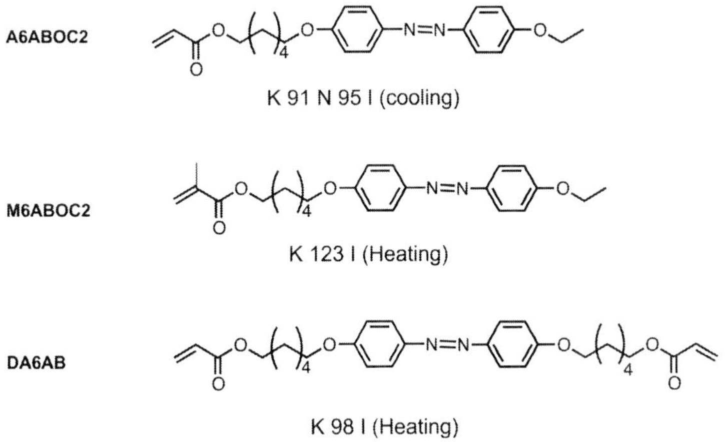

A6ABOC2

K 91 N 95 I (cooling)

M6ABOC2

K 123 I (Heating)

DA6AB

K 98 I (Heating)

Figure 2.5 Thermal properties of azobenzene-containing monomers. Acylate (above), methacrylate (middle), and crosslinker (below).

2.1.4 Lateral Substituents

In general, the presence of lateral substituents will broaden the LC molecules, thus reducing lateral attractions and lowering stabilities of nematic and smectic phases. With an addition of a substituent ortho to the polar terminal group, it can hinder the molecular association and leads to a higher dielectric anisotropy ($\Delta\varepsilon$) (Fig. 2.6).

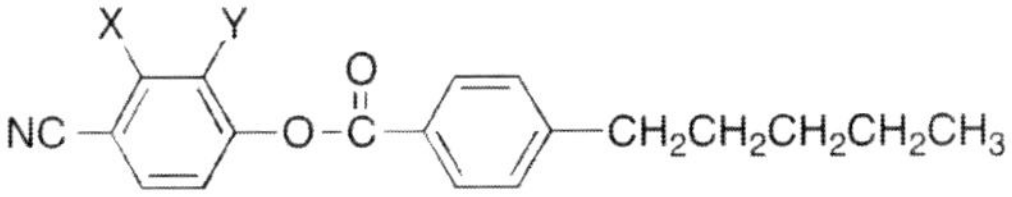

X	Y	C–N/°C	N–I/°C	$\Delta\varepsilon$
H	H	56.8	63.4	20.7
H	F	20	30	48.9

Figure 2.6 Influence of the lateral group on LC properties (C, crystal; N, nematic; I, isotropic phase).

2.2 Some Physical Parameters

In LC materials, liquid-like positional order exists at least in one direction of space. Some degree of long-range ordering is induced by orientation of molecules or molecular aggregates. LCs have mechanical properties of liquids: high fluidity, inability to support shear, and formation of droplets. However, one of the most important properties of LC is anisotropy in optical, electrical, magnetic, and possibly in mechanical properties.

2.2.1 Order Parameter

The structure of LC phases is usually characterized by a "so-called" director (n), which is a unit vector, showing an average orientation of the molecular axes in some macroscopic bulk. The amount of orientational order in LCs can be obtained by averaging macroscopically molecular orientation with respect to this direction. This order parameter (S) can be evaluated by measuring the average angle (θ) between the director and the long axes of the mesogens (Fig. 2.7). For an isotropic fluid, $S = 0$, whereas for a perfectly aligned crystal, $S = 1$. For typical LCs, S is between 0.3 and 0.8, and this value generally decreases due to higher mobility and disorder as the temperature is raised. The S sharply decreases at the clearing point or phase transition temperature, as shown in the right of Fig. 2.7.

2.2.2 Mechanical Properties (Elastic Constant)

When LC materials are deformed under an external field, they tend to restore their original states. To describe their mechanical properties, three elastic constants, splay (K_{11}), twist (K_{22}), and bend (K_{33}) are defined according to the elastic theory (Fig. 2.8). These parameters determine the relaxation speed of LC devices. For a twisted nematic (TN) LCD, a small bend to splay elastic constants (K_{33}/K_{11}) is favorable. Molecules with aromatic systems and long alkyl chain often give a small K_{33}. Generally, the order of magnitude of elastic constant is about 10^{-11} N. Figure 2.8 also gives some commercially available LC examples and their typical applications from electronic materials (EM) industry.

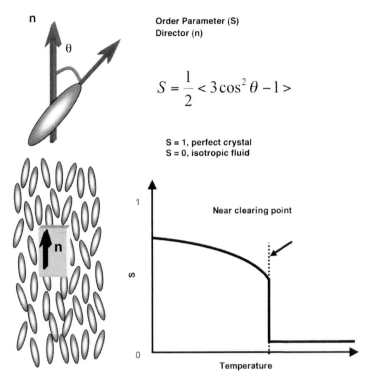

Figure 2.7 The geometrical structure used for defining the order parameter (S) of the nematic LC phase. The angle θ denotes the deviation of the long axis of one individual mesogen from the director.

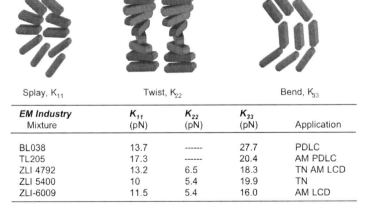

EM Industry Mixture	K_{11} (pN)	K_{22} (pN)	K_{33} (pN)	Application
BL038	13.7	------	27.7	PDLC
TL205	17.3	------	20.4	AM PDLC
ZLI 4792	13.2	6.5	18.3	TN AM LCD
ZLI 5400	10	5.4	19.9	TN
ZLI-6009	11.5	5.4	16.0	AM LCD

Figure 2.8 Three elastic constants of LC materials.

2.2.3 Electrical Properties (Dielectric Anisotropy)

As a result of the uniaxial anisotropy, an electric field experiences a different dielectric constant when oscillating in a direction parallel or perpendicular to the director (n). The differences, $\Delta\varepsilon = \varepsilon_\parallel - \varepsilon_\perp$ is called the dielectric anisotropy. If the dielectric constant along the director ($\varepsilon_\parallel$), is larger than that in the direction perpendicular to it ($\varepsilon_\perp$), one calls it positive anisotropy (Fig. 2.9). Under an electric field, LCs with positive anisotropy orient with their molecular long axis along the electric direction (E), whereas LCs with negative anisotropy orient perpendicularly to E. The magnitude of $\Delta\varepsilon$ determines the strength of interaction between the LCs and applied electric field. A larger $\Delta\varepsilon$ may endow LC devices with a short turn-on time and a low threshold voltage.

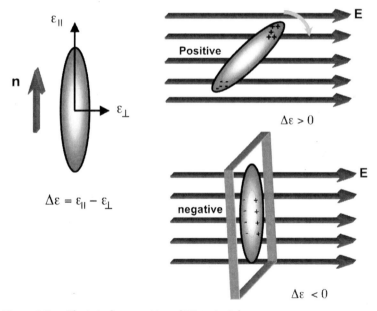

Figure 2.9 Electrical properties of LC materials.

2.2.4 Magnetic Anisotropy (Diamagnetism)

In general, most of LC materials are magnetic anisotropy (or diamagnetic), which originates from their dispersed electron distributions associated with the electron structures of mesogens.

The magnetic field (H) causes a molecular current, which produces a magnetic field opposing the applied field. Therefore, the delocalized charge makes the major contribution to diamagnetism. As shown in Fig. 2.10, the magnetic susceptibility tensor is defined as χ. Ring currents associated with aromatic units give a large negative component to χ for a direction perpendicular to aromatic ring plane. $\Delta\chi$ is usually positive, since $\Delta\chi = \chi_\parallel - \chi_\perp > 0$.

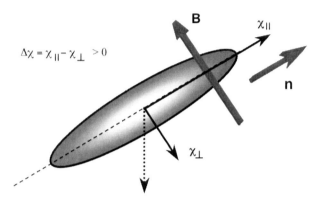

Figure 2.10 Schematic illustration of magnetic anisotropy of LC materials.

2.2.5 Optical Anisotropy (Birefringence)

Liquid crystal materials show two principal refractive indices: extraordinary refractive index (n_e) and ordinary refractive index (n_o), as shown in Fig. 2.11. Optical anisotropy (or birefringence, $\Delta n = n_e - n_o$) is related to the anisotropic polarizability of LC molecules, which is greatly dependent of the extent of conjugation of π-bonds. LC molecules containing aromatic rings and unsaturated bonds often exhibit a higher birefringence. Besides, the value of birefringence is also dependent on the wavelength of the probe light and the temperature of measurement. As shown on in Fig. 2.11, it decreases with increase in the measurement temperature [Balzarini, 1970].

When light with a wavelength of λ passes through an LC samples with a thickness of d, phase retardation (or phase shift $\Delta n \bullet d$) occurs. The condition $n_e > n_o$ describes a positive uniaxial LC, so that nematic LCs are in this category. For typical nematic LCs, n_o is approximately 1.5 and their birefringence may range between 0.05 and 0.5.

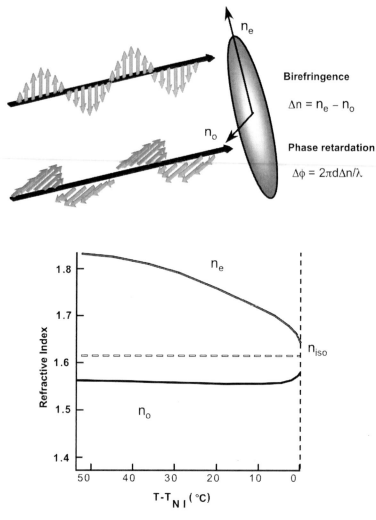

Figure 2.11 Optical anisotropy of LC materials (above) and dependence of LC refractive index on temperature (below).

2.2.6 Fréedericksz Transition

In 1927, Fréedericksz and Repiewa first reported that phase transition occurred when a sufficiently strong electric field (higher than E) is applied to LC materials in an undistorted state [Fréedericksz & Repiewa, 1927]. This does not occur at the surface of the LC molecule, due to strong interactions within the substrate molecules that keeps them fixed (Fig. 2.12). Reorientation takes place in the rest of the LC molecules along the field direction. The Fréedericksz transition is fundamental to the operation of many LCDs because the director orientation (and thus, the properties) can be controlled easily by the application of a field.

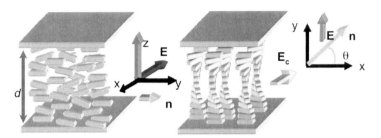

Figure 2.12 Electrically controlled Fréedericksz transition. The director rotates at some critical E field. Before E_c nothing happens.

2.3 LC Alignment

The LC materials possess both an ordering of crystalline materials and easy controllability of an isotropic liquid, which enables their molecular orientation to be easily manipulated. LC molecules are easily and strongly responsive to various types of external stimuli such as electric or magnetic fields, or the presence of surfaces with particular properties. Thus, the LC order can often be extended over macroscopic scales. In many cases, one can align the LC molecules along any selected directions. Moreover, since the hierarchical structure appears in an LC phase, it is generally possible to get the system to rearrange by simply applying a new external field, resulting in the change of physical properties. The response can be as fast as microseconds and strong enough to be utilized in a whole range of applications. Figure 2.13 shows two kinds of LC alignment, in-plane and out-of-plane ways according to the positions of LC direction and the substrate plane. Among them, homogeneous and homeotropic alignment is generally used.

During the development of LC devices, several types of alignment layers and aligning methods have been proposed and investigated. They can roughly be categorized by the nature of the material employed, organic or inorganic. On the other hand, contact and non-contact methods are included according to the treatment ways for alignment layers.

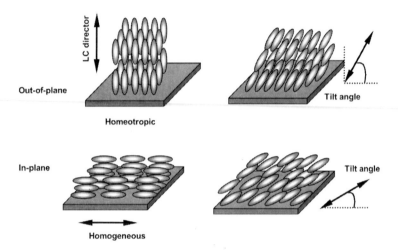

Figure 2.13 In-plane and out-of-plane alignment of LCs.

2.3.1 Rubbing Technique

The simplest method to get planar orientation of LCs is mechanical rubbing of the surface of a glass pre-coated with an alignment layer, as shown in Fig. 2.14. Under the unidirectionally rubbing function with a cotton or velvet cloth mounted on a cylinder, using a special rubbing machine, microgroove structures are often produced in the form of ridges and troughs, which promotes the orientation of the LC molecules along these formations. The pressure, moving speed and cylinder diameter of the roller are all important process parameters in determining the LC alignment.

Surprisingly, the rubbing process first applied in the late 1930s to achieve LC alignment still dominates the process commercially applied for assembly of LCDs today [Zöcher, 1925]. After the rubbing treatment, aligned LC molecules often show an angle between their directors and the rubbing direction, which is called pre-tilt angle (Fig. 2.14). It has been well known that pre-tilt angle prevents the creation of reverse tilt disclinations in LCDs. For homeotropic alignment, specific polyimides are available (e.g., Nissan SE-1211).

Without the rubbing process, perfect homeotropic alignment is obtained.

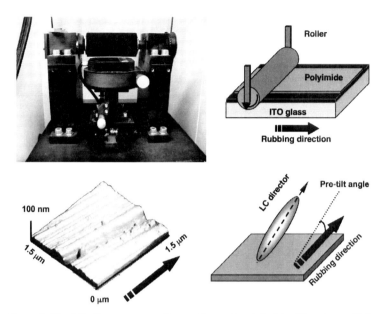

Figure 2.14 LC alignment on the surface of polyimide films by rubbing treatment. Picture of a rubbing machine and the scheme of rubbing process (above). Atom force microscopy (AFM) image showing a rubbed polyimide film coated on a glass slide (below).

Two mechanisms have been proposed to explain the rubbing mechanism, which has been debated for decades. For thermosetting polymers (like polyimide), microgrooves with high regularity and suitable interval (about 30 nm) appear along the rubbing direction after the rubbing process. It is the surface deformation that induces alignment of LCs. On the other hand, no microgroove structures form on the surface of thermoplastic polymers like, polyamide, poly(vinyl alcohol) (PVA), and polyester. But a statistically important number of similarly oriented bonds accumulate on the surface of the thermoplastic polymer by the rubbing process [Rasing & Musevic, 2004; Takatoh et al., 2005]. In this case, intermolecular interactions between the LCs and the molecules of the alignment layer are responsible for the LC alignment. However, the detailed mechanism still remains unclear.

The mechanically rubbing technique has the advantage of being relatively straightforward, easy, and cheap, and generates very stable alignment layers. As it is already a widely applied process in industry, no large investments are required. Unfortunately, dust or static electricity might be produced, which could induce defects to the macroscopic alignment. Moreover, this contact method can only be applied on a flat surface; it shows no function on a curve one. Therefore, many new approaches for LC alignment have been developed afterwards.

2.3.2 Microgroove Method

Following the microgroove mechanism of rubbing technique, other methods have been proposed to produce microgroove structures on the surface of alignment layers. For examples, surface-relief gratings (SRGs) inscribed in photoresponsive polymer films by interference of two coherent laser beams can be used to control LC alignment, as shown in Fig. 2.15. LC molecules orientate along the surface-relief direction because the system energy is the lowest at this state. In addition to SRGs, soft lithography can be also used to generate microgrooves on glass substrates. More interestingly, nano-imprinted lithography on plastic substrate can avoid alignment layer for assembling soft LCDs [Lin et al., 2009]. Besides, modification of polymer surface has been obtained with mechanically scratching with a tip from a scanning probe microscope (SPM). Although the microgroove method for LC alignment induced a lower pre-tilt angle, it undoubtedly facilitates researches into the exact nature of alignment mechanism.

Figure 2.15 Schemes of LC molecules are aligned along the microgroove direction.

2.3.3 Electric and Magnetic Fields

As described in Section 2.2, LC molecules show dielectric anisotropy and diamagnetism. Such anisotropic properties are responsible for their alignment under electric and magnetic fields. As shown in Fig. 2.16, LCs with positive electrical anisotropy orient with their directors along the electric direction (E), whereas LCs with negative electrical anisotropy orientate perpendicularly to E. Similarly, LCs can be aligned along the magnetic direction.

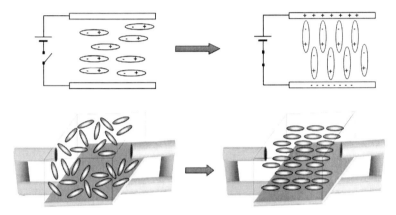

Figure 2.16 LC alignment by an electric field (above) and a magnetic field (below).

2.3.4 Surfactant Dipping and LB Membrane

Some surfactants like lecithin and silicone compounds can be used to chemically modify glass surface, endowing the treated substrates with capability of anchoring LCs. For instance, a lecithin-treated surface can induce homeotropic alignment of LCs, as shown in Fig. 2.17. Using silicone compounds with a long alkyl chain as active agents can bring the substrate a similar function of LC alignment. Generally, such an alignment layer show more stable than that treated with lecithin.

Langmuir–Blodgett (LB) films are ultrathin layers of organic materials that have been deposited with a special technique, since LB membrane often shows ordered structure formed in the assembly process, as shown in Fig. 2.18. Generally, to form polyimide

LB films, the polyamic acids should be first changed into ammonium salts, which are used for assembly of LB films. Then, the obtained LB film of polyamide acid is converted into polyimide by thermal or by acidic treatments. Some theoretical work has already been developed [Barbero & Petrov, 1994], but the application of this somewhat exotic technique is in practice restricted to laboratory spaces.

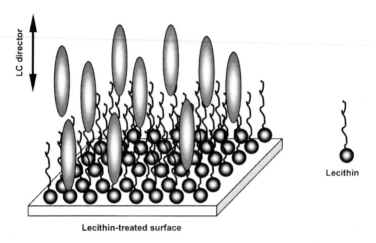

Figure 2.17 LC alignment on the lecithin-treated surface of one glass slide.

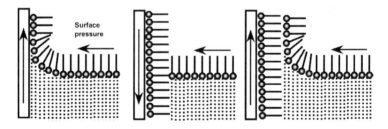

Figure 2.18 Fabrication of LB films for LC alignment.

Gold-sputtered plates can be covered with a self-assembly monolayer of chemically absorbed alkanethiols [Gupta & Abbott, 1997]. Experiments revealed an odd–even effect of the length of the alkyl tail on the alignment properties. When odd alkanethiols were used, the LC molecules would align parallel to the surface, but perpendicular to the direction of the deposition of the gold layer (Fig. 2.19a). In contrast, when even-numbered alkyl tails were used,

LC molecules would align parallel to both the surface and deposition direction of the gold layer (Fig. 2.19b). When a mixture of even- and odd-numbered tails was used, the LC molecules adopted a homeotropic orientation (Fig. 2.19c).

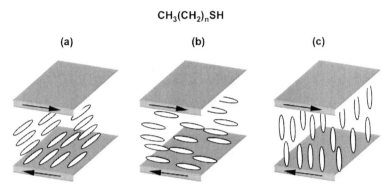

Figure 2.19 LC alignment on self-assembly monolayers of chemically absorbed alkanethiols: (a) perpendicular to the direction of the deposition of the gold layer, (b) parallel to both the surface and deposition direction of the gold layer, and (c) homeotropic orientation.

2.3.5 Supramolecular Self-Assembly

In 2003, a self-assembly method was reported to control LC alignment [Hoogboom et al., 2003], as shown in Fig. 2.20. When an indium tin oxide (ITO) plate was immersed into a solution of the naphthalene-functionalized siloxane, a highly ordered surface was formed by supramolecular self-assembly. Such a film was capable of inducing uniform LC alignment, which could be attributed to two simultaneously occurred processes. One is related to the creation of stronger intermolecular interactions between the large π-surfaces of naphthalene-functionalized siloxane and the other involve the covalent grafting of these precipitating oligosiloxanes onto the ITO surface.

2.3.6 Oblique Evaporation

Oblique evaporation of inorganic alignment layers such as silicon oxide was found to induce LC alignment since 1977 [Pollack et al.,

1977]. The LC alignment effect is correlated with the deposition angle, as shown in Fig. 2.21. The principal difficulty with conventional evaporation techniques for depositing alignment layers onto obliquely orientated substrates is that of maintaining a uniform angular distribution of depositing atoms at the substrate and hence, a uniformly deposited layer. Both homogeneous and homeotropic LC alignment can be easily obtained by controlling the evaporation direction, as shown in Fig. 2.22.

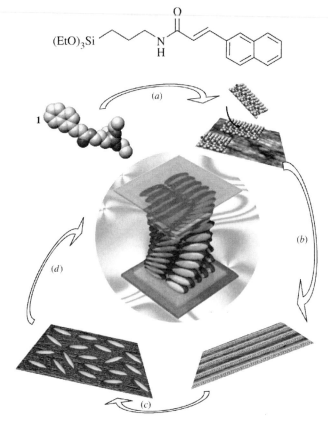

Figure 2.20 LC alignment based on supramolecular self-assembly of naphthalene functionalized siloxane. Reprinted with permission from Hoogboom et al., Copyright 2002, WILEY-VCH Verlag GmbH & Co. KGaA, Weinheim.

Advantages of the oblique evaporation technique include the fact that it is a contactless method, posing no risk of damaging the substrate either mechanically or by electrostatic discharge. The deposited layers are thinner than their organic equivalents, so less influence on the optical and dielectric behavior of the device is to be expected. Furthermore, the superior stability of inorganic layers removes all concerns about lifetime and reliability of the alignment system.

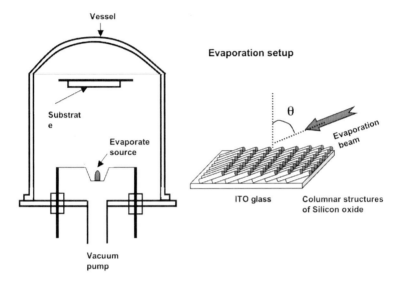

Figure 2.21 Evaporation setup for preparation of LC alignment layers.

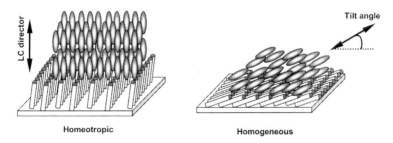

Figure 2.22 Homogeneous and homeotropic LC alignment on evaporated layers.

However, a considerable disadvantage may be the cost of the vacuum equipment. The needed directionality of the beam limits the size of the substrates or calls for large vacuum chambers, both considered unfavorable for high volume production. This can be circumvented with some engineering effort, for example, the above-mentioned beam shaping or the use of linear evaporation sources. It is also fair to say that some deposition techniques that have been proposed are quite complex and do not always feature an equally high reproducibility.

2.3.7 Ionic and Plasma Beams

The use of ion-beam irradiation for LC alignment can be traced back to around 1979 [Little et al., 1979], and the use of plasma-beam irradiation to 1986 [Kurchatkin et al., 1986]. In 2001, IBM deposited a diamond-like carbon hard coat with 3 to 15 nm thick via chemical vapor deposition or sputtering [Chaudhari, 2001], whose surface was bombarded with argon-ion beam under an oblique incidence angle to prepare alignment layers, showing a uniform homogeneous alignment of LC molecules and strong anchoring energy.

Very similarly, plasma beams can produce alignment layers by modification of chemical bonds on surface. The movement of a plasma source breaks the aromatic rings and creates dangling carbon bonds that subsequently form directional C–O bonds on the surface, which are thought to be responsible for LC alignment [Gwag et al., 2004]. Although the mechanism is still unclear, the bombardment of substrate surfaces with energetic particles from plasma or ion beam sources can lead to changes in the surface topography from ion/particle surface interactions.

2.4 Photoalignment

As described in Section 2.3.1, LCs can be aligned along the rubbing direction on polyimide films because of the lower energy level along the unidirectional rubbing-induced grooves, which has been widely applied in the present LCD industry. But the rubbing method has some obvious demerits such as, the contaminated dust and static charge produced during the rubbing process, which may cause serious problems like, the speckles and reduced display contrast. Moreover, the rubbing method can only be applied on flat surfaces, which has no function on curve ones. Therefore, the non-contact method of photoalignment is considered to be one of the promising ways to solve the above-mentioned problems and can be easily used to assemble LC displays with high quality, wide viewing angles, and large-area screen. According to mechanism of photochemical reactions, the photoalignment method for alignment layers can be divided into photoisomerization, photocrosslinking, and photodegradation.

2.4.1 Photoisomerization (Command Surface)

The photoalignment of LCs with self-assembled monolayers of azobenzene dyes formed on glass substrates was first reported by silane coupling agents [Ichimura, 2000]. Figure 2.23 shows schematic illustration of the basic concept of the photoalignment technique, which is also called "command surfaces". In LC cells made from the surface-treated substrates, nematic LCs exhibit homeotropic alignment when azobenzenes immobilized on the substrates are in the transform. Photoirradiation to cause *trans–cis* isomerization of azobenzenes enables repeatable changes in alignment from homeotropic to planar states. Similarly, photoalignment change of LCs in-plane can be acquired by the "command surfaces" via photoalignment of azobenzenes in plane.

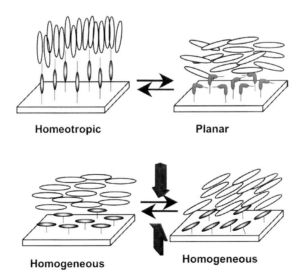

Figure 2.23 LC alignment on command surface (or photoisomerization).

Besides, the photoalignment of LCs has been also reported for polymer films containing azobenzene moieties, the effects of structure and density of azobenzenes on the photoalignment behavior have been systematically studied. The photoalignment function of azobenzene dyes is not limited on the surface, and a similar effect was found in volume of polymer films doped with azobenzene molecules. As shown in Fig. 2.24, an azobenzene-doped polyimide film exhibits capability of control alignment of LCs by linearly polarized light (LPL) [Gibbons, 1991]. The direction of the homogeneous alignment of LC molecules in cells could be altered by changing the polarization direction of the actinic light.

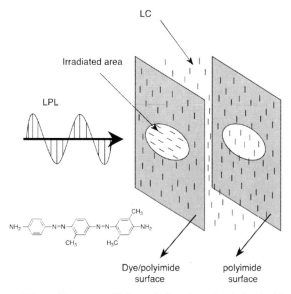

Figure 2.24 Photoalignment of LCs with dye-doped polyimide films.

A nematic LC cell fabricated from one substrate coated with an azobenzene/polyimide mixture and the other substrate coated only with polyimide, with the rubbing directions of both substrates mutually parallel, was exposed to LPL with polarization parallel to the rubbing direction. It was revealed that the alignment of LC molecules at the irradiated area of dye-doped surface changed from rubbing direction to perpendicular to the polarization of the actinic light, whereas those at the undoped polyimide surface remained unchanged, resulting in a TN structure within the irradiated region.

The photoalignment can be subsequently erased or rewritten by altering the polarization direction of LPL to induce a change in azobenzene alignment. Furthermore, the unidirectional alignment of LCs has been also achieved by PVA thin films containing an azobenzene dye.

Then many photoisomerizable materials containing azobenzenes have been developed as the photoalignment layers. Figure 2.25 gives two typical examples, in which azobenzene was designed in the side chain and main chain of polymers, respectively. In most cases, LCs are oriented perpendicularly to the polarization of the LPL since LC molecules preferentially affected by *trans*-azobenzenes. The thermal stability of the photoalignment was enhanced when azobenzenes is input as one part of main chain in polyimide, as shown in Fig. 2.25. On the other hand, main-chain type materials also showed a higher stability because of their higher glass-transition temperature.

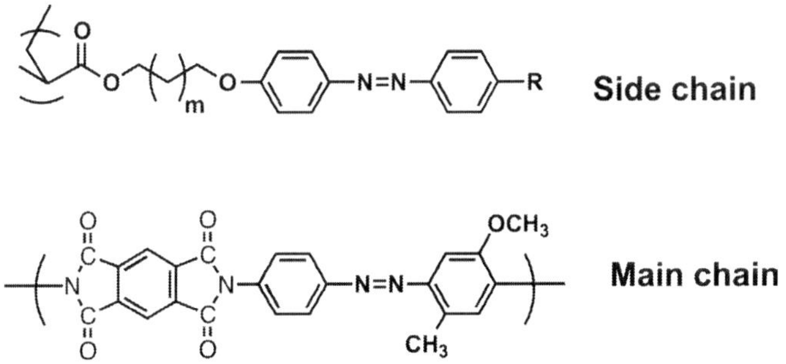

Side chain

Main chain

Figure 2.25 Two examples of photoisomerizable materials for LC alignment.

2.4.2 Photocrosslinking (or Linearly Photodimerized) Reaction

Different from photoisomerization of azobenzene dyes, a novel photoalignment method was developed based on anisotropic photocrosslinking of optically active molecules [Schadt et al., 1992]. Figure 2.26 shows illustration of photocrosslinking method of linearly photodimerized reaction for alignment layers using poly(vinyl cinnamate), in which (2+2) photodimerization of cinnamates in the side chain occurs anisotropically, leading to optical dichroism in films. The formed cyclobutane derivatives and hydrocarbon chains result in anisotropically dispersive surface interaction forces with LC molecules, which is responsible for the homogeneous alignment of LCs, perpendicularly to the polarization direction of the irradiated linearly polarized UV light (LPUV) [Schadt et al., 1992]. Such a photoalignment method of non-contact way using LPUV is more favorable from the viewpoint of improvement of performance of the existing LC devices and development of novel LC devices with new functions, because the photoalignment process is free from dust particles and static surface charges generated in the conventional rubbing method involving a contact process.

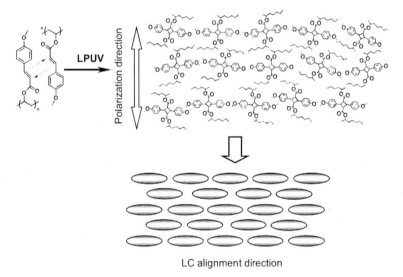

Figure 2.26 Scheme of anisotropic photocrosslinking method for LC alignment.

Other photosensitive polymers like epoxy, polyimide, and polyurethane with cinnamates in side chain also showed a similar photoalignment behavior of poly(vinyl cinnamate) [Yu et al., 2003a, 2003b, 2004; Xia et al., 2005]. Figure 2.27 summarizes some materials containing photocrosslinkable cinnamates for LC alignment upon anisotropic photocrosslinking reaction. The LC alignment behaviors of them were based on the axis-selective photoreaction of the cinnamte groups. In addition to polymethacrylates, epoxy-based materials shown in Fig. 2.27 also possessed good alignment capacity for LCs. Similarly, to the photoisomerizable materials, polyimide-based photocrosslinkable materials exhibited good thermal stability for the LC photoalignment. But such a kind of photoalignment layer demonstrates a low energy of azimuthal anchoring compared with rubbed polyimide alignment layers, which is widely used in assembly of commercially available LCDs.

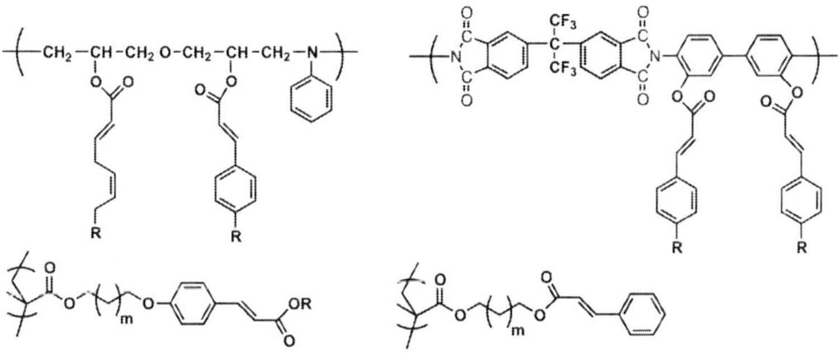

Figure 2.27 Molecular schemes of materials containing photocrosslinkable cinnamate moieties for LC alignment.

To improve the capability of LC photoalignment in axis-selectively light-reactive polymeric films with cinnamate groups, various factors have been considered to determine the orientation direction of LCs. If the photoreactive moieties align perpendicularly to the LPL polarization direction and preferentially interact with LCs as indicated by Schadt [Schadt et al., 1992], the LC alignment direction is perpendicular to LPL polarization because LCs align along the photoreactive moieties. That is to say, the photoproducts weaken the interaction with LCs, LCs are oriented along the unreacted cinnamate direction, resulting in perpendicular LC alignment. Contrastively, when the direction of the photoreactive moieties is parallel to LPL polarization direction and the photoproducts preferentially interact with LCs, LC alignment parallel to the polarization direction is achieved.

Apart from cinnamate-containing polymers, other materials with chromophores such as coumarin [Schadt et al., 1996], tolane [Obi et al., 1999], chalcone [Yu et al., 2002a, 2002b, 2003c], stilbene, styrylpyridine, phenylacetylene, and anthracene groups have also been explored for their photoalignment behaviors based on the LPUV technique. LC photoalignment by means of anisotropic photocrosslinking was also found in polyesters containing a phenylenediacryloyl group in the main chain, and good alignment effects were found when the aliphatic spacer in the main chain unit had an odd number of methylene repeated units. To improve the anisotropy of alignment films formed by LPUV, Kawatsuki et al. directly attached cinnamate to a biphenyl mesogen as a side chain of photosensitive polymers, and the LC alignment was greatly improved because of the enhanced molecular reorientation upon annealing [Kawatsuki et al., 1997]. Another advantage of this photoalignment method is the easy fabrication of multi-domain LC alignment for high-quality displays [Schadt et al., 1996].

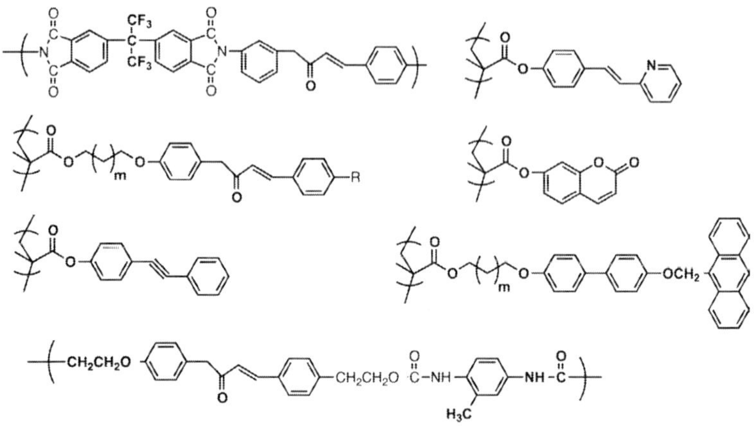

Figure 2.28 Several materials containing photocrosslinkable chromophores other than cinnamtes for photoalignment layers via LPUV method.

2.4.3 Photodegradation (Photoactive Polyimide)

The main problem hindering commercial exploitation of photoalignment layers is their poor thermal stability; therefore much attention was not paid to polyimide, which is well known for its high thermal stability of LC alignment. Then, a unique LC alignment technique based on a novel principle was proposed: a dynamic and static control of LCs by means of photoactive polyimides as an alignment layer. The most significant feature of this system is to control alignment of LCs without photochemical reactions. In 1995, the first photoalignment layers were reported upon exposure of polyimide films at 257 nm [Hasegawa & Taira, 1995]. However, this method is accompanied by chemical reactions, resulting in weakening the thermal stability of polyimides due to serious degradation of backbone structures.

To improve the thermal stability of polyimides further, a benzophenone moiety was used to design a photoactive polyimide as an alignment layer, and a dynamic photocontrol of alignment of LCs, optical switching of nematic LCs were obtained [Kim et al., 1999, 2001]. As shown in Fig. 2.29, an LC cell with a gap of 5 µm, in which both substrates were coated with rubbed polyimide films, was fabricated to evaluate the photoresponse of a nematic LC (5CB). The change in transmittance as a function of applied voltage (electric Fréedericksz transition) for the LC cell is shown in the left of Fig. 2.29, in which the threshold voltage for the electric Fréedericksz transition is quite different between the irradiated and unirradiated samples. It was found that photoirradiation at 366 nm results in an immediate change in transmittance when a bias voltage of 4.5 V is applied across the LC cell. Since the transmittance recovers when photoirradiation is stopped, it is assumed that the optical switching observed is based on an alignment change between a homogeneous state and a homeotropic state. Although the switching mechanism is not well-understood, formation of intra- or intermolecular charge transfer complexes in the surface area of polyimide films upon photoirradiation (i.e., a change in polarity) seems to participate in the optical switching.

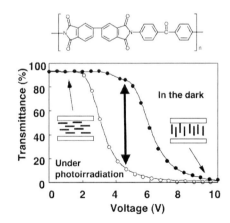

Figure 2.29 Electric-optical effect of a 5CB cell with photoresponsive polyimide as an alignment layer. Reprinted with permission from Yu and Ikeda, Copyright 2011, WILEY-VCH Verlag GmbH & Co. KGaA, Weinheim.

Recently, aromatic polyimides exposed to LPUV light with a longer wavelength were reported to show a high efficiency of photoalignment without any significant change in chemical structures of polyimides [Wang et al., 1998a, 1998b, 1998c]. To evaluate the alignment ability, a polyimide film formed on a glass substrate was irradiated with LPUV at 366 nm, and then an LC cell with a gap of 5 μm was fabricated with two pieces of these exposed polyimide-coated substrates, sandwiching with 5CB between them. It has been demonstrated that unidirectionally homogeneous alignment of 5CB is successfully induced with the photoactive polyimide film. The degree of uniformity of LC alignment is strongly influenced by the used polyimide structures and it has been shown that an aromatic polyimide with a unit of diphenyl ether diamine is more favorable as a photoalignment layer in terms of photosensitivity and chemical stability [Wang et al., 2001]. The alignment layer prepared by exposure to LPUV after completing imidization of poly(amide acid) (PAA) films enables LC alignment perpendicular to the polarization direction of LPUV. However, if LPUV exposure is carried out during the process of thermal imidization of PAA, alignment of LC molecules was induced parallel to the polarization direction of LPUV, as shown in the lower part of Fig. 2.30 [Xu et al., 2003]. This approach is referred to as the in situ method, which might restrain the degradation process since the photoirradiation and thermally induced imide cyclization were performed simultaneously. With the same structure of a photoactive polyimide, two different LC alignment directions can be conveniently achieved upon exposure to LPUV merely by changing the thermal treatment process.

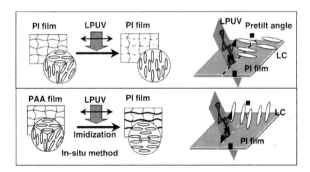

Figure 2.30 Photoalignment based on degradable polymides. PI, polyimide. Reprinted with permission from Yu and Ikeda, Copyright 2011, WILEY-VCH Verlag GmbH & Co. KGaA, Weinheim.

References

Balzarini, D. A. (1970). Temperature dependence of birefringence in liquid crystals, *Phys. Rev. Lett.* **25**, pp. 914–915.

Barbero, G. and Petrov, A. G. (1994). Nematic liquid crystal anchoring on Langmuir-Blodgett films: Steric, biphilic, dielectric and flexoelectric aspects and instabilities, *J. Phys. Condens. Matter* **6**, pp. 2291–2306.

Chaudhari, P., Lacey, J., Doyle, J., Galligan, E., Lien, A., Callegari, A., Hougham, G., Lang, N., Andry, P. S., John, R., Yang, K. H., Lu, M., Cai, C., Speidell, J., Purushothaman, S., Ritsko, J., Samant, M., Sthr, J., Nakagawa, Y., Katoh, Y., Saitoh, Y., Sakai, K., Satoh, H., Odahara, S., Nakano, H., Nakagaki, J., and Shiota. Y. (2001). Atomic-beam alignment of inorganic materials for liquid-crystal displays, *Nature* **411**, pp. 56–59.

Fréedericksz, V. and Repiewa, A. (1927). Theoretisches und experimentelles zur Frage nach der natur der anisotropen flüssigkeiten, *Zeitschrift für Physik Society* **42**, pp. 532–546.

Gibbons, W. M., Shannon, P. J., Sun, S. T., and Swetlin, B. J. (1991). Surface-mediated alignment of nematic liquid crystals with polarized laser light, *Nature* **351**, pp. 49–50.

Gupta, V. K. and Abbott, N. L. (1997). Design of surfaces for patterned alignment of liquid crystals on planar and curved substrates, *Science* **276**, pp. 1533–1536.

Gwag, J. S., Jhun, C. G., Kim, J. C., Yoon, T. H., Lee, G. D., Cho, S. J. (2004), Alignment of liquid crystal on a polyimide surface exposed to an Ar ion beam, *J. Appl. Phys.* **96**, pp. 257–260.

Hasegawa, M. and Taira, Y. (1995). Nematic homogeneous photo alignment by polyimide exposure to linearly polarized UV, *J. Photopoly. Sci. Technol.* **8**, pp. 241–248.

Hoogboom, J., Behdani, M., Elemans, J. A. A. W., Devillers, M. A. C., De Gelder, R., Rowan, A. E., Rasing, T., and Nolte, R. J. M. (2003). Noncontact liquid-crystal alignment by supramolecular amplification of nanogrooves, *Angew. Chem. Int. Ed. Eng.* **42**, pp. 1812–1815.

Little, M. J., Garvin, H. L., and Lee, Y. S. (1979). Means and method for inducing uniform parallel alignment of liquid crystal material in a liquid crystal cell, US Patent 4153529.

Ichimura, K. (2000). Photoalignment of liquid-crystal aystems, *Chem. Rev.* **100**, pp. 1847–1874.

Kawatsuki, N., Ono, H., Takatsuka, H., Yamamoto, T., and Sangen, O. (1997). Liquid crystal alignment on photoreactive side-chain liquid crystalline polymer generated by linearly polarized UV light, *Macromolecules* **30**, pp. 6680–6682.

Kim, G. H., Enomoto, S., Kanazawa, A., Shiono, T., Ikeda, T., and Park, L. S. (1999). Optical switching of nematic liquid crystal by means of photoresponsive polyimides as an alignment layer, *Appl. Phys. Lett.* **75**, pp. 3458–3460.

Kim, G. H., Enomoto, S., Kanazawa, A., Shiono, T., Ikeda, T., and Park, L. S. (2001). Application of photosensitive polyimides as alignment layer to optical switching devices of nematic liquid crystal, *Liq. Cryst.* **28**, pp. 271–277.

Kurchatkin, S., Mauraveva, N. A., Mamaev, A. L., Sevostyanov, V. P., and Smirnova, E. I. (1996). Method of producing orienting layer of liquid crystal indicator. RU2055384C1.

Lin, T., Yu, S., Chen, P., Chi, K., Pan, H., and Chao, C. (2009). Fabrication of alignment layer free flexible liquid crystal cells using thermal nanoimprint lithography, *Curr. Appl. Phys.* **9**, pp. 610–612.

Obi, M., Morino, S., and Ichimura, K. (1999). Photocontrol of liquid crystal alignment by polymethacrylates with diphenylacetylene side chains, *Chem. Mater.* **11**, pp. 1293–1301.

Pollack, J. M., Haas, W. E., and Adams, J. E. (1977). Topology of obliquely coated silicon monoxide layers, *J. Appl. Phys.* **48**, pp. 831–833.

Rasing, T. and Musevic, I. (2004). *Surfaces and interfaces of liquid crystals*, Berlin, Heidelberg, Germany: Springer-Verlag.

Schadt, M., Schmitt, K., Kozinkov, V., and Chigrinov, V. (1992). Surface-induced parallel alignment of liquid crystals by linearly polymerized photopolymers, *Jpn. J. Appl. Phys. Pt. 1* **31**, pp. 2155–2164.

Schadt, M., Seiberle, H., and Schuster, A. (1996). Optical patterning of multi-domain liquid-crystal displays with wide viewing angles, *Nature* **381**, pp. 212–215.

Takatoh, K., Hasegawa, M., Koden, M., Itoh, N., and Hasegawa, R. (2005). *Alignment technologies and applications of liquid crystals*. London, UK: CRC Press.

Wang, Y., Xu, C., Kanazawa, A., Shiono, T., Ikeda, T., Matsuki, Y., and Takeuchi, Y. (1998a). Generation of nematic liquid crystal alignment with polyimides exposed to linearly polarized light of long wavelength, *J. Appl. Phys.* **84**, pp. 181–188.

Wang, Y., Xu, C., Kanazawa, A., Shiono, T., Ikeda, T., Matsuki, Y., and Takeuchi, Y. (1998b). Alignment of a nematic liquid crystal induced by anisotropic photooxidation of photosensitive polyimide films, *J. Appl. Phys.* **84**, pp. 4573–4578.

Wang, Y., Kanazawa, A., Shiono, T., Ikeda, T., Matsuki, Y., and Takeuchi, Y. (1998c). Homogeneous alignment of nematic liquid crystal induced by polyimide exposed to linearly polarized light, *Appl. Phys. Lett.* **72**, pp. 545–547.

Wang, Y., Xu, C., Kanazawa, A., Shiono, T., Ikeda, T., Matsuki, Y., and Takeuchi, Y. (2001). Thermal stability of alignment of a nematic liquid crystal induced by polyimide exposed to linearly polarized light, *Liq. Cryst.* **28**, pp. 473–475.

Xu, C., Shiono, T., Ikeda, T., Wang, Y., and Takeuchi, Y. (2003). Photo-induced alignment of liquid crystals on polyimide parallel to polarization of linearly polarized light, *J. Mater. Chem.* **13**, pp. 669–671.

Yu, H. F., Bai, H., Lian, Y., and Wang, X. (2002a). One photoalignment layer materials of polyurethane for display and its preparation method. CN 01129616.X, 1325936.

Yu, H. F., Bai, H., Lian, Y., and Wang, X. (2002b). Photoalignment layer materials containing chalcone in main chain for display and their preparation method. CN 01129214.8, 1322792.

Yu, H. F., Bai, H., Lian, Y., Wang, X., and Liu, D. (2003a). Synthesis and characterization of expoxy-based photoalignment layer materials. *Chem. J. Chin. Univ.* **24**, pp. 165–168.

Yu, H. F., Jiang, H., Lian, Y., Wang, X., and Liu, D. (2003b). Synthesis and characterization of polyurethane-based photo-alignment layers materials, *Acta. Polym. Sin.* **1**, pp. 133–138.

Yu, H. F., Jiang, H., Lian, Y., and Wang, X. (2003c). Synthesis and characterization of polyurethane-based photo-alignment layers materials for LCDs, *Polymer Prepr.* **224**, pp. 618–619.

Yu, H. F., Zhang, M., Lian, Y., and Wang, X. (2004). Study on polyimide-based photoalignment layer materials, *Acta. Polym. Sin.* **1**, pp. 22–26.

Xia, Q., Yu, H. F., Lian, Y., and Wang, X. (2005). Cinnamate-containing polyimide and its function as liquid crystal photo-alignment layer, *Acta. Polym. Sin.* **6**, pp. 914–918.

Zöcher, H. (1925). Optical anisotropy of selectively absorbing substances; mechanical production of anisotropy, *Naturwissenschaften* **13**, pp. 1015–1021.

Chapter 3

Light and Liquid Crystals

Liquid crystals (LCs) are easily and strongly responsive to various external stimuli such as electric, magnetic, optical fields, and so on. Combining the unique electro-optical properties of LCs with the alignment feature with the help of rubbed polyimide films, electric response of LCs has been widely applied in production of commercially available flat-panel displays. Therefore, LCs have become a household name as a result of their widespread applications in television, smartphone, and computer displays.

As shown in Fig. 3.1, a low applied voltage reorients the low-molecular-weight (LMW) LC molecules from the homogeneous state induced by the rubbed polyimide substrates to the vertical alignment state. By this way, a display element can be fabricated by changing the optical properties under an electric field, and bright and dark states are achieved accordingly. On removing the electric field, the LCs relaxes back to the substrate-induced ground state. The response can be as fast as microseconds and as strong enough to be applicable in a whole range of applications.

Light is one of the most important clean energies, which can be regarded as limitless comparing with other energy forms. Furthermore, its easy control and processability enable it to deeply influence the development of LC science, especially when photochromic molecules are introduced into LC systems. During the past two decades, photochromic molecules are attracting the growing interest as guests in the field of materials science due to their ability

---

Dancing with Light: Advances in Photofunctional Liquid-Crystalline Materials
Haifeng Yu
Copyright © 2015 Pan Stanford Publishing Pte. Ltd.
ISBN 978-981-4411-11-0 (Hardcover), 978-981-4411-12-7 (eBook)
www.panstanford.com

to change the properties of the host system, such as its electronic or ionic conductivity, fluorescence, magnetism, and shape, upon irradiation with light of the appropriate wavelength. Combining the photosensitivity of photochromic molecules and self-organization of LCs (Fig. 3.2), photoresponsive LCs show unique performance which will be introduced in this chapter. On the one hand, photoinert LCs can be controlled by the photochemical reactions of photosensitive materials. On the other hand, photoresponsive LCs can be directly orientated by light when photochromic groups act as mesogens.

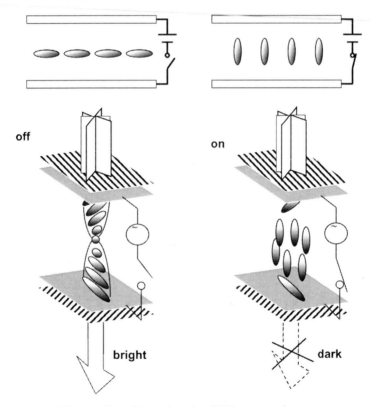

Figure 3.1 Scheme of working principle of LCDs.

3.1 Photochemical Reactions

As discussed in Chapter 2, LC molecules can be uniaxially oriented on thin polymer films with physical (typically optical) anisotropy.

To achieve such anisotropic properties, the photoreaction should be carried out in an anisotropic way. As far as simple and convenient ways are concerned, photoisomerization and photocrosslinking are often chosen. Their typical representatives are azobenzene and cinnamates, as shown in Fig. 3.3.

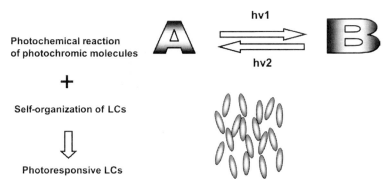

Figure 3.2 Photoresponsive LCs combine the photosensitivity of photochromic molecules and self-organization of LCs.

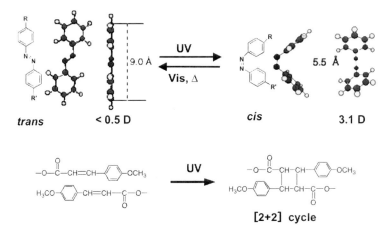

Figure 3.3 Scheme of photoisomerization (above) and photocrosslinking (below).

In the field of photoresponsive LCs, azobenzene is one of the most important moieties, which has been extensively studied. Firstly, the synthetic way of azobenzene groups is well established. The most popular preparation method of azo coupling reaction often exhibits a high yield (Scheme 3.1). Furthermore, diverse of molecular structures are easily tailored, whereas heavy synthesis work has been often omitted. A myriad of azobenzene derivatives can be tailored by modifying the substituents in the two benzene rings, resulting in their maximum absorption from the ultraviolet (UV) to visible (vis) regions.

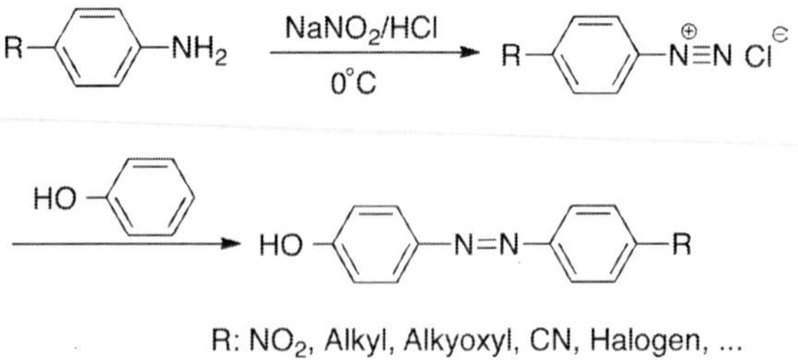

Scheme 3.1 One example of azo coupling reaction for preparation of photoisomerizable azobenzene compounds.

Secondly, its photoisomerization possessed highly sensitive and almost completely reversible with little fatigue. Figure 3.4 gives an example of UV–vis absorption spectra of one azobenzene compound in solution. Upon UV irradiation, the absorption at 350 nm due to the π–π^* transition decreased quickly, whereas another absorption at 450 nm due to the n–π^* transition gradually increased [Yu et al., 2005]. Two isobestic points at 315 nm and 410 nm were observed, indicating that only one kind of photoreaction (photoisomerization) occurred. Furthermore, this photoresponse occurs only in one single molecule, which makes it seldom interfered by external environment.

Thirdly, *trans*-to-*cis* or its back isomerization often induces a large change in molecular conformation, which is also accompanied with a change in transition moment (Fig. 3.3). This leads to their far different effect on LC molecules in photoresponsive systems, which is described in Fig. 3.5. *Trans*-azobenzene is rod-like and symmetric in molecular configuration, which generally stabilizes an LC phase. On the contrary, the molecular shape of *cis*-azobenzene is bend and soft, showing a high transition moment, which results in destabilization of LC phases.

Fourthly, azobenzene itself may act as a mesogen and self-organize into diverse of LC phases, which results in the marriage of photochromic groups with LCs. Photoinduced LC-to-isotropic phase transition can be directly obtained within a range of femi-seconds. Upon molecular or supramolecular cooperative motions (SMCM), the nanoscale deformation of photoisomerization can be magnified into macroscopic scale. Therefore, most of the photoresponse of LCs in this book is mainly based on azobenzene-containing LC systems.

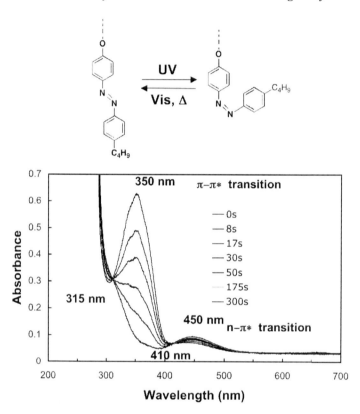

Figure 3.4 Photoisomerization of one typical azobenzene materials in solution.

Similarly, photocrosslinking or (2+2) photo cycloaddition reaction has also been studied for their photocontrol capability of LC molecules. This photoresponse needs participation of two photochromic molecules, different from the photoisomerization of azobenzene. Besides, other groups such as stilbenes, imines, spiropyranes, fulgides, and diarylethenes, have also explored for such properties.

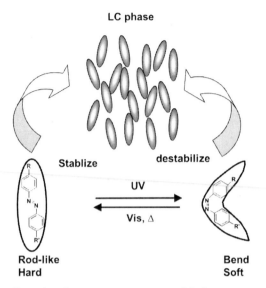

Figure 3.5 Effect of azobenzene isomers on an LC phase.

3.2 Photoinduced Alignment and Reorientation of LCs

On the surface of polymer thin film, uniaxial alignment of LCs has been achieved due to the anisotropic interaction between the alignment-layer film and rod-like LC molecules. Generally, alignment layers for control of LMW LCs need only a small anisotropy when axis-selective photoreaction occurs. But things are far different for optical device as far as practical applications are concerned, which often requires a large birefringence and a thick film. Moreover, photoinduced molecular reorientation and molecular cooperative motion (MCM) are beneficial for this purpose. In this case, the photochromic molecule may act as both a photoresponsive moiety and a mesogen. In the following section, the photoresponsive behaviors of such a kind of LC materials will be discussed.

3.2.1 Photoresponse to Linearly Polarized Light

Upon light irradiation, molecules can transmit from their basic states to excited states. But such photoabsorption should meet exclusive requirements including the coincidence of the electric field vector of actinic light with the direction of transition moments of photochromic molecules (photoselection). Azobenzene is one of typical photochromic molecules. Its *trans*-isomers show π–π^* transition moments approximately parallel to the molecular long axis, enabling them to angular-dependent absorption of the actinic light. Photoalignment occurs upon irradiation of one beam of linearly polarized light (LPL), as shown in Fig. 3.6.

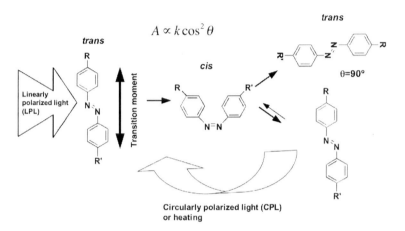

Figure 3.6 Photoalignment of azobenzene molecules.

Upon photoirradiation, azobenzenes with their transition moments parallel to the polarization direction of LPL are effectively activated to their excited states, followed by *trans–cis* isomerization in solutions or solid states, leading to a large change in molecular shape and polarizability. The probability of the absorption is proportional to the $\cos^2\theta$ (Fig. 3.6), where θ is the angle between the transition moment of an azobenzene and the LPL polarization direction. Other molecules with their transition moments perpendicular to the LPL polarization are inactive toward isomerization. The *cis*-isomers can return to their *trans*-isomers by thermal treatment or visible light irradiation. Combining the polarization-selective *trans*-to-*cis* isomerization and unselective back *cis*-to-*trans* isomerization, the number of azobenzene moieties with their transition moments normal to the light polarization gradually increases. This means that a net population of *trans*-azobenzenes aligned perpendicularly to the light polarization can be obtained at the end of the multi-cycles of repetition of *trans–cis–trans* isomerization. Thus, light-selective photoalignment is achieved, which is well known as the "Weigert effect" [Weigert, 1919]. By this way, azobenzenes can be reorientated to any controlled directions by choosing appropriate polarization direction of light [Yu & Ikeda, 2011]. As shown in Fig. 3.6, the ordered alignment state can be deoriented by circularly polarized light (CPL) or thermal treatment.

Figure 3.7 gives one example of photoalignment of azobenzene-containing LC polymer (LCP) in film with one LPL beam. Here, an LCP showing smectic LC phase was used. The spin-coated LCP film was put between a pair of two crossed polarizers. An LPL laser beam at 488 nm was used to induce LC alignment. A He–Ne laser at 633 nm with weak intensity was adopted as a probe beam since no absorption at this wavelength was observed in the UV–vis absorption spectrum of the LCP film. The transmittance as a probe beam was measured simultaneously during irradiation through the two crossed polarizers with the sample film between them.

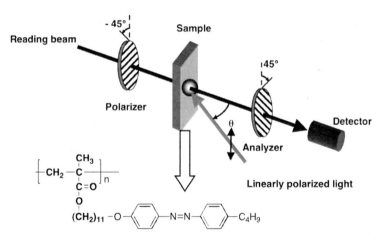

Figure 3.7 Experimental setup for photoalignment of azobenzene-containing LCP film.

The experimental results are given in Fig. 3.8. Before irradiation, the transparent film showed isotropy since no alignment of the mesogens was obtained by film formation process. On irradiation, the alignment of azobenzene moieties was first induced, leading to a quick initial increase in transmittance (T). After 20 s irradiation, T gradually increased. Then, the curve leveled to a saturated value at 100 s because almost all the azobenzene mesogens were aligned. No decrease in T was observed after turning off the laser beam, indicating that a stable photoinduced birefringence was obtained by the photoinduced LC alignment. Generally, T is defined by $T = \sin^2(\pi d \Delta n/\lambda)$, where d is the film thickness, Δn is the photoinduced birefringence, and λ is the wavelength of the probe light.

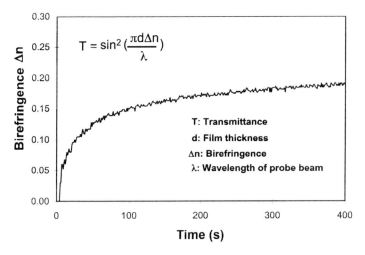

Figure 3.8 Experimental results of photoalignment of azobenzene-containing LCP film in Fig. 3.7.

After photoalignment, the irradiated film exhibited an intensive anisotropy in its polarized UV–vis spectra (Fig. 3.9). The order parameter (S) of about 0.2 at 350 nm is calculated by $S = (A_\perp - A_\parallel)/(A_\perp + 2A_\parallel)$, where $A_\perp$ and $A_\parallel$ are the absorbance perpendicular and parallel to the polarization direction of the actinic laser beam, respectively. This indicates that the alignment of the LCs is perpendicular to the polarization direction of the actinic LPL.

3.2.2 Thermal Effect on Photoalignment

Large photoinduced anisotropy can be obtained when the polarization-selective photoreaction is accompanied by molecular reorientation. Moreover, thermal enhancement of the photoinduced optical anisotropy of the photoresponsive LCP also plays an important role in assembling LC devices. For instance, polymethacrylate with 4-methoxyazobenzene side groups shows thermally enhanced in-plane or out-of-plane molecular reorientation, depending on the length of the alkylene spacer and the amount of the axis-selectively formed *cis*-isomers, as illustrated in Fig. 3.10. LCPs with a short spacer produced thermal enhancement of in-plane homogenous alignment. When the spacer is longer enough (with more than eight carbon atoms), out-of-plane orientation was observed.

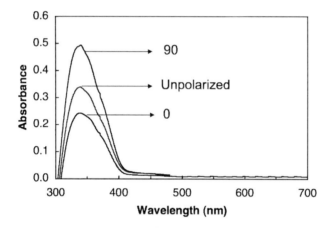

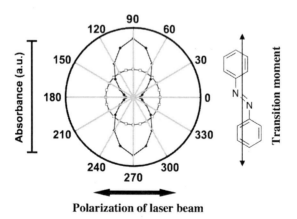

Figure 3.9 Polarized UV–vis spectra of the LCP film after photoalignment with LPL.

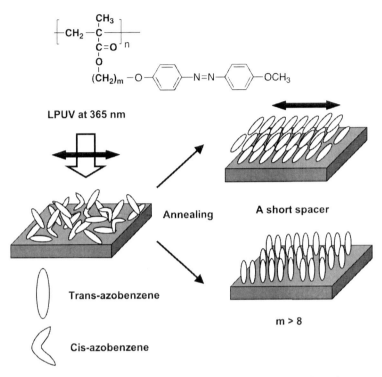

Figure 3.10 Photoalignment and thermal enhancement of azobenzene LCPs with different spacers in the side chain.

The reason why the out-of-plane orientation was obtained upon thermal annealing still remains challengeable. However, LCPs showing a smectic phase are more easily inclined to demonstrate this thermal effect after being annealed at the smectic phase. It was reported that a polyacrylate copolymer having both 4-ethoxy-4′-hexyloxy azobenzene and cholesterol groups as a side-chain group showed spontaneous out-of-plane molecular orientation on a glass substrate upon annealing at 75°C, whereas less out-of-plane molecular orientation was observed for a polyacrylate homopolymer having the same azobenzene side group [Bobrovsky et al., 2004]. In contrast, it was found that a polyacrylate homopolymer consisting of 4-methoxy-4′-hexyloxy azobenzene aligned spontaneously in the out-of-plane orientation on a glass substrate upon annealing at a smectic phase [Moritsugu et al., 2011]. It is interesting to note that a very small difference in the terminal substituents has a large effect on the spontaneous molecular orientation behaviors of the LCPs. Figure 3.11 shows one example of one smectic LCP upon annealing. The absorption at about 350 nm due to the $n-\pi^*$ transition of azobenzenes decreased greatly after annealing, because of the out-of-plane orientation of the mesogens.

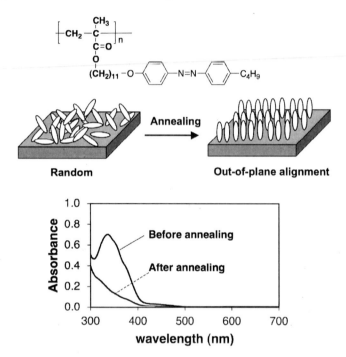

Figure 3.11 Schematic illustration and UV–vis absorption spectra of azobenzene-containing LCP thin films before and after being annealed at its smectic phase, without any light irradiation.

Several groups have investigated photoinduced reorientation parallel to polarization of azobenzene-containing LCP films. This technique uses pretreatment of non-polarized UV light and then photoalignment, as shown in Fig. 3.12. The *trans*-azobenzene groups are almost off resonant at the longer wavelength of the actinic light. Molecular reorientation is based on an axis-selective photoreaction of the *cis*-isomers, which are prepared by exposure to non-polarized UV light upon pretreatment. Kempe et al. have reported the photoinduced reorientation of polymeric films with 4-cyanoazobenzene side groups using LPL at 633 nm combined with pre-exposure to unpolarized UV light to create a photostationary state of the *cis*-isomers prior to irradiating with 633 nm light [Kempe et al., 2003; Zebger et al., 2003]. As shown in Fig. 3.12, the thermally enhanced reorientation of LCs using polarized light at 633 nm has achieved an in-plane reorientational order greater than 0.8 [Uchida et al., 2004, 2005]. In this case, the orientation behavior is independent of the alkylene spacer length.

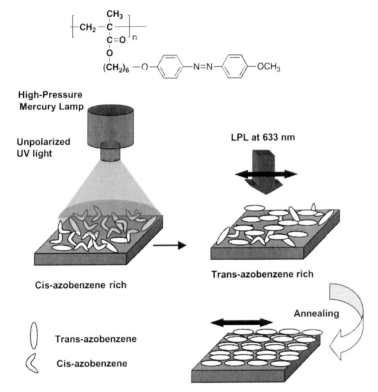

Figure 3.12 Photoalignment and thermal enhancement of azobenzene LCPs upon pretreatment. The ordered state is parallel to the polarization direction of LPL at 633 nm.

3.2.3 Thermal Enhancement of Photoalignment in Photocrosslinkable LCs

Kawatsuki et al. first reported the thermally enhanced effect on the photoinduced molecular reorientation in LCPs containing photocrosslinkable cinnamate or cinnamic acid as side groups [Kawatsuki et al., 2002; Uchida & Kawatsuki, 2006]. As shown in Fig. 3.13, the photocrosslinking reaction occurs only when both of the transition moments of cinnamates parallel to the polarization direction of the linearly polarized UV (LPUV) light, and no reaction for the cinnamate perpendicular to the light polarization. That is to say, an axis-selective photocrosslinking often occurs when a photocrosslinkable LCP film is irradiated with LPUV light. However, the mesogens are confined in its glassy state at room temperature, and molecular reorientation does not carry out, resulting in a small photoinduced optical anisotropy by the anisotropic crosslinking without thermal treatment. Heating the small anisotropic films up to their LC temperature but below the clearing point, MCM occurs, which leads to molecular reorientation of the unreacted mesogens. Such effect often produce strong enhancement of the photoinduced small anisotropy of the photoresponsive LCPs.

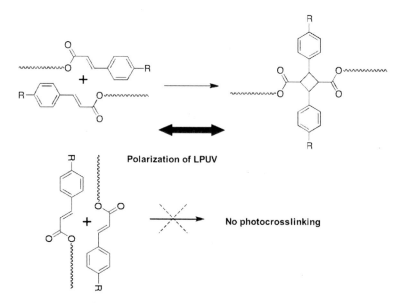

Figure 3.13 Scheme of anisotropic photocrosslinking reaction in cinnamate-containing LCPs upon irradiation with LPUV.

The thermally induced molecular reorientation can be classified into two cases, according to the degree of anisotropic photoreaction. On one hand, irradiation with a low dose of actinic light, a low crosslinking degree is obtained accordingly and the axis-selectively formed products act as impurities. In such a film, a thermal treatment at the LC temperature produces molecular reorientation with their transition moment perpendicular to the polarization of the LPUV light (the left route of Fig. 3.14). This situation is similar to the thermal enhancement of azobenzene-containing polymers in Fig. 3.10. On the other hand, when a high crosslinking

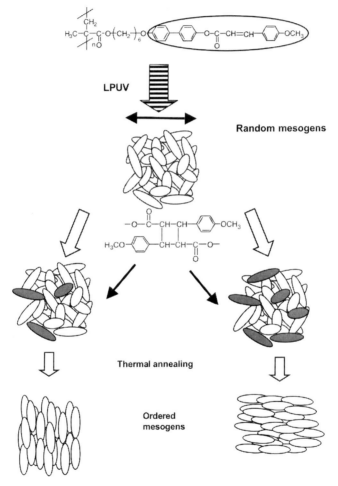

Figure 3.14 Photoalignment and thermal enhancement of photocrosslinking LCPs. Reprinted with permission from Yu and Ikeda, Copyright 2011, WILEY-VCH Verlag GmbH & Co. KGaA, Weinheim.

degree is induced, the thermal annealing can generate enhanced molecular reorientation with their transition moment parallel to the polarization of the light. In this case, the photocrosslinked products by the axis-selectively photoreaction act as an alignment anchor to thermally reorient the non-reacted mesogens (the right portion of Fig. 3.12) [Kawatsuki et al., 2002]. If the degree of photocrosslinking is two high, most of the mesogens are confined by the crosslinked network, thermal annealing only bring about little effect on the molecular reorientation.

Since the photocrosslinkable LCPs showing a large photoinduced anisotropy was first reported [Kawatsuki et al., 2002], several other photocrosslinkable LCPs have been developed to generate in-plane photoinduced molecular orientation combined with thermal treatment (Scheme 3.2). Without thermal annealing at LC temperature, only a very small optical anisotropy is generated after the anisotropic photoreaction. Generally, the obtained order parameter (S) is smaller than 0.1 and the photoinduced birefringence (Δn) is smaller than 0.01. Upon annealing the LPUV-irradiated LCP films at the LC temperature, the induced molecular reorientation can results in greatly enhanced optical anisotropy ($S > 0.5$, $\Delta n > 0.15$).

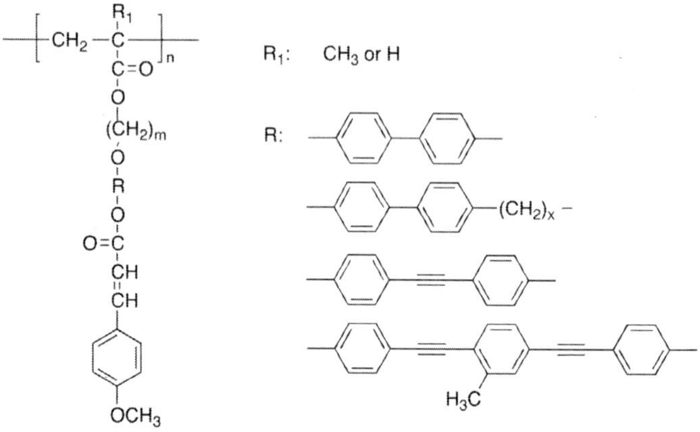

Scheme 3.2 Some examples of photocrosslinkable LCPs for LC alignment.

Recently, tolane groups have been incorporated with the photoinduced anisotropic system. A photoinduced Δn, as large as 0.5, was obtained in tolane-containing photoresponsive LCPs because of the large inherent birefringence of the tolane moieties than that of other mesogens [Okano et al., 2006a, 2006b]. Similarly, photocrosslinking LCPs containing tolane mesogenic groups also exhibit a large photoinduced Δn upon molecular reorientation [Kawatsuki et al., 2008, 2010].

As shown in Fig. 3.15, when photocrosslinkable LCPs are slantwise irradiated with LPUV, 3D molecular reorientation is observed parallel to the polarization direction of LPUV upon annealing [Kawatsuki et al., 2001]. Different from azobenzene-containing LCPs generally exhibiting yellow color, cinnamate derivatives are *trans*-parent in the visible region, especially for cinnamate-containing LCPs, which promise them to be more useful for display applications [Kawatsuki et al., 2001].

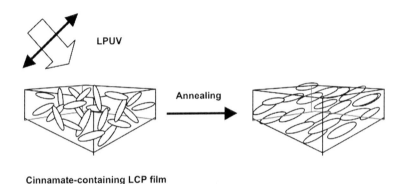

Cinnamate-containing LCP film

Figure 3.15 Three-dimensional (3D) (out-of-plane) photoalignment of photocrosslinking LCPs.

3.2.4 Photoresponse to Unpolarized Light

When azobenzene groups possess soft alkyl substituents in their benzene rings, they often self-assemble into an ordered LC phase, as shown in Fig. 3.16. At this case, azobenzene moieties play both roles of photoresponsive groups and rod-like mesogens. Upon irradiation of one beam of unpolarized light, the well-known *trans*-to-*cis* photoisomerization occurs, which is accompanied by the photoinduced LC-to-isotropic phase transition, since the *trans*-azobenzene can be a mesogen and the *cis*-azobenzene never shows any LC phase due to its bent shape. Generally, the photochemical reaction of azobenzenes is reversible; *cis*-to-*trans* back isomerization can be caused upon irradiation with visible light or thermal treatment, resulting in an isotropic-to-LC phase transition. Thus, light can be easily applied to control the changes between the ordered LC and disordered isotropic states. This is very important for the photonic applications of photoresponsive LC materials, which will be introduced in Chapter 5.

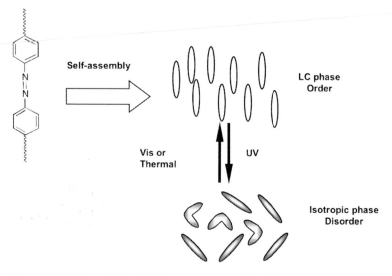

Figure 3.16 Photoinduced phase transition of azobenzene-containing LC materials.

The alignment direction of azobenzene-containing polymers can be controlled in a 2D manner (in-plane) by changing the polarization direction of the actinic light. In the research of in-plane alignment process, photoinduced biaxiality of azobenzene moieties was observed in LC and amorphous polymer films and in Langmuir–Blodgett films [Schönhoff et al., 1996]. This phenomenon was interpreted in terms of the realignment of azobenzene moieties along the propagation direction of the light (the out-of-plane alignment).

According to the Weigert effect [Weigert, 1919], only the azobenzene mesogens with their transition moments perpendicular to the polarization direction of the actinic light is inactive. When an unpolarized UV light is used to irradiate samples, the azobenzenes along the propagation direction of the light is inactive and photoisomerization occurs with difficulty, because the propagation direction of light is always perpendicular to its electric field vector, as shown in Fig. 3.17. Azobenzene moieties in other directions should undergo *trans–cis–trans* cycles and the final state should be in the inactive direction. That is, to say, the photoisomerizable azobenzenes are aligned along the incident direction of the light. By this way, 3D photoalignment (out-of-plane) is easily obtained.

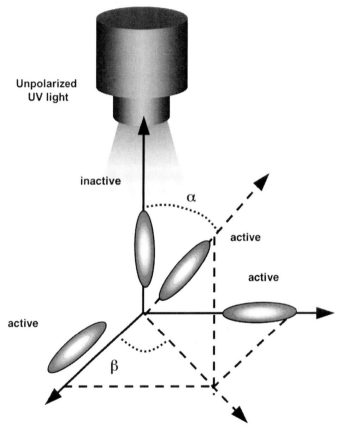

Figure 3.17 Photoinduced 3D alignment of azobenzene LCs upon irradiation with unpolarized light.

In fact, several works have been reported concerning the out-of-plane alignment behavior by using unpolarized light. From the viewpoint of 3D manipulation of molecules by light, interestingly promising results have been obtained for photoresponsive LCPs.

It is easy to photocontrol alignment of azobenzene and mesogenic moieties in a 3D fashion, precisely along the propagation direction of the actinic light at room temperature, by changing the incident angle of the actinic unpolarized light, as shown in Fig. 3.18 [Wu et al., 1999b]. Furthermore, it has been also confirmed that the induced 3D alignment of molecules is very stable below T_g of the LCPs. These results are expected to open a new methodology to photo-manipulate molecules in a 3D manner.

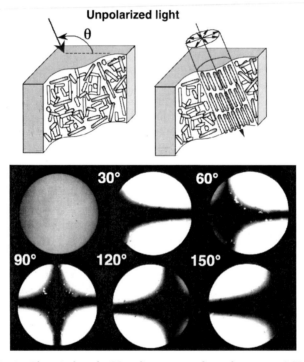

Figure 3.18 Photoinduced 3D alignment of azobenzene LCPs with unpolarized UV light. The LC alignment direction is along the incident direction of the light. Reprinted with permission from Wu et al., Copyright 1999, WILEY-VCH Verlag GmbH, Weinheim, Fed. Rep. of Germany.

Here, it must be emphasized that the photoinduced 3D alignment of photoisomerizable azobenzene LC materials is obtained by irradiation with unpolarized light. For the photocrosslinkable cinnamate-containing LC materials, their 3D alignment is achieved by slantwise irradiation with LPUV, as shown in Fig. 3.15. The comparison between the photoisomerizable and photocrosslinkable LCs upon photoirradiation is summarized in Table 3.1.

Table 3.1 Summary of photoisomerizable and photocrosslinkable LCs upon photoirradiation

Photoresponsive LCs	Photoisomerizable LCs	Photocrosslinkable LCs
Typical representatives	Azobenzene	Cinnamate
Appearance	Yellow	Transparent in visible light
Photochemical reaction	Isomerization Phase transition	(2+2) crosslinking isomerization
In-plane photoalignment	LPL	LPL with thermal treatment
Out-of-plane photoalignment	Unpolarized light	LPL with thermal treatment
Anisotropy induced by photoreaction	Big	Small
Thermal enhancement after photoreaction	Small	Big
MCM	Yes	Yes

3.3 Photomodulation of LCs

As one of green and neat energies, light is particularly fascinating since it supplies precise and reversible controls with a non-contact way. The photoinduced alignment and molecular orientation can be precisely induced upon irradiation of light-responsive LCs containing photoisomerizable or photocrosslinkable photochromic groups, which enables one to photomodulate them into hierarchical patterns by adjusting the actinic light with wavelength, intensity, polarization, phase retardation, interference, and so on. On the one hand, photoresponsive LCs have been widely utilized as good light actuators for beam steering, phase retardation, wavefront modulation, and polarization switching and control. On the other hand, the change in LC alignment can also be reflected by its optical properties. Both of them are the bases for photonic applications of LC materials.

Figure 3.19 shows several ways of photocontrol of light-responsive LCs materials. In LMW LCs and LCPs, the photomodulation among three states of LC alignment can be easily obtained by choosing suitable input actinic light. Detecting with LPL, LC materials show different refractive index of n_e (extraordinary refractive index), n (refractive index in an isotropic phase), n_o (ordinary refractive index), respectively, when they are in parallel and perpendicular alignment (ordered LC states) and in a random isotropic phase (a disordered state).

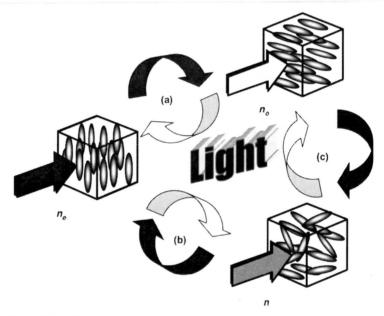

Figure 3.19 Photomodulation of LC properties (a) is an order–order change, (b) and (c) are order–disorder changes (also phase transition changes). Reprinted with permission from Yu and Ikeda, Copyright 2011, WILEY-VCH Verlag GmbH & Co. KGaA, Weinheim.

As shown in Fig. 3.19a, LPL with its electric vector oscillating in the vertical direction detects n_e, when LC molecules are in vertical alignment. Whereas, n_o can be measured by LPL if the LC molecules change their alignment from perpendicular to parallel orientated state. The change in refractive index in Fig. 3.19a corresponds to the LC birefringence, $\Delta n (= n_e - n_o)$. The birefringence of LC materials is often quite large, typically 0.1~0.2, so that photoinduced change in

LC alignment usually brings about a large change in refractive index. Different from the order–order change in Fig. 3.19a–c are order–disorder changes, which are correlated with alterations of refractive index (n_e–n) and (n–n_o), respectively. The order of the refractive index is usually $n_e > n > n_o$, thereby the change in the LC directors (order–order change) gives rise to a larger change in refractive index than the order–disorder changes, although the order–disorder changes still give a large alteration of refractive index comparing with amorphous materials.

In LC elastomers (LCEs), LC molecules cannot be singly photomodulated because they are 3D crosslinked to form network structures, enabling them to be macroscopically induced change in shape by external stimuli. de Gennes et al. [1997] first theoretically proposed thermally induced uniaxial contraction of LCEs in the direction of the optic axis or director axis by combination the anisotropic aspects of LC phases and the rubber elasticity of polymer networks. This was then developed and experimentally obtained by Finkelmann et al. [1981]. A large change in volume of LCEs promises them to be explored as novel materials for artificial muscles and soft actuators (Fig. 3.20).

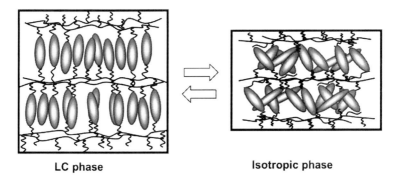

LC phase   Isotropic phase

Figure 3.20 Scheme of artificial muscles and soft actuators based on LCEs.

Similarly, the deformation of LC elastomers (LCEs) can be induced by actinic light, as shown in Fig. 3.21, bringing about photomechanical and photomobile properties of LCEs. Upon UV irradiation, photomodulation of LCs occurs only in the surface region of LCE films because of the large molar extinction coefficient of photochromic LCs and light cannot penetrate through a thick film (e.g., thickness >10 µm) which induces the volume change only in film surface and generates photomechanical effect, as shown in Fig. 3.21.

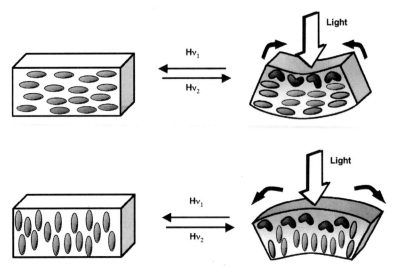

Figure 3.21 Photomechanical effect in LCEs. The homogenous alignment of LCs produces bending toward the actinic light (above) and the homeotropic alignment of LCs causes bending away from the light source (below). Reprinted with permission from Yu and Ikeda, Copyright 2011, WILEY-VCH Verlag GmbH & Co. KGaA, Weinheim.

When the alignment of LC molecules is parallel to the surface of substrates, volume contraction is produced just along the pre-aligned direction, contributing to the anisotropic bending behavior toward the incident direction of the actinic light source. On the contrary, volume expansion is brought about when the LC molecules are aligned perpendicularly to the substrates, which results in different bending behaviors, away from the actinic light source. Furthermore, the photomechanical behaviors are reversible if azobenzene molecules are used as photoresponsive mesogens [Ikeda et al., 2003]. By this way, light energy can be directly transferred into mechanical energy, which attracts much attention of scientists in photoresponsive LCEs and makes it become one of hot spots in the present LC research.

In LC block copolymers (LCBCs) with at least one mesogenic polymer chain as one of constituent blocks, LC molecules in the continuous substrates of bulk films can be used as ordering actuators to control the alignment of microphase-separated nanostructures by SMCM, as shown in Fig. 3.22. By the interplay processes between regular periodicity of LC ordering and thermally controlled microphase separation, the ordering of mesogens can be transferred to the nanostructures inside them, leading to periodically ordered nanostructures with regular packing in macroscopic scale. The orientation of the hierarchically assembled nanomaterials strongly correlated with the alignment direction of the photoresponsive LC molecules [Yu et al., 2006a, 2006b], which makes the microphase-separated nanostructures in LCBCs superior to those in amorphous ones since the LC alignment can be easily achieved in an arbitrary scale with already developed photoalignment techniques. The supramolecularly assembled nanostructures in LCBC films show excellent reproducibility and mass production, which provides reliable templates for nanofabrication processes, leading to widely applications in macromolecular engineering.

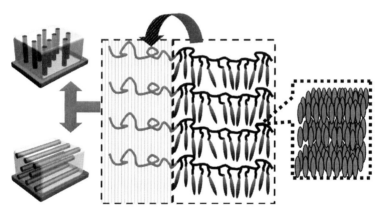

Figure 3.22 SMCM in LCBC films is responsible for the formation of microphase-separated nanostructures with a macroscopic ordering. Reprinted with permission from Yu and Ikeda, Copyright 2011, WILEY-VCH Verlag GmbH & Co. KGaA, Weinheim.

The engrossing photoresponsive LC materials offer an effective and convenient chance to adjust properties of advanced materials by integrating photoresponsive molecules with LC properties, which has become one of the emerging topics in the present and future LC researches. Hence, it is the right time to compile a comprehensive textbook on the nature of this type of materials and the ongoing progress in this research area, which will be useful to professionals working in this field. The goals of this book are to summarize previous work, provide new insights into this class of photoresponsive LC materials, and add to the understanding of the potential applications in flat panel display, photonics, data storage, photomobile devices, and nanotechnology.

References

Bobrovsky, A., Boiko, N., Shibaev, V., and Stumpe, J. (2004) Comparative study of photoorientation phenomena in photosensitive azobenzene-containing homopolymers and copolymers, *J. Photochem. Photobiol. A: Chem.* **163**, pp. 347–358.

de Gennes, P. G. (1975). Physique moleculaire, *C. R. Acad. Sci. B* 281, pp. 101–103.

de Gennes, P. G., Hebert, M., and Kant, R. (1997). Articifical muscles based on nematic gels, *Macromol. Symp.* **113**, pp. 39–49.

Finkelmann, H., Kock, H. J., and Rehage, H. (1981). Investigations on liquid crystalline polysiloxanes. 3. Liquid crystalline elastomers-a new type of liquid crystalline material, *Makromol. Chem. Rapid Commun.* **2**, pp. 317–322.

Han, M. and Ichimura, K. (2001). In-plane and tilt reorientation of p-methoxyazobenzene side chains tethered to liquid crystalline polymethacrylates by irradiation with 365 nm light, *Macromolecules* **34**, pp. 90–98.

Ikeda, T., Nakano, M. Yu, Y., Tsutsumi, O., and Kanazawa, A. (2003). Anisotropic bending and unbending behavior of azobenzene liquid-crystalline gels by light, *Adv. Mater.* **15**, pp. 201–205.

Kawatsuki, N., Goto, K., Kawakami, T., and Yamamoto, T. (2002). Reversion of alignment direction in the thermally enhanced photoorientation of photo-cross-linkable polymer liquid crystal films, *Macromolecules* **35**, pp. 706–713.

Kawatsuki, N., Kawakami, T., and Yamamoto, T. (2001). A photoinduced birefringent film with a high orientational order obtained from a novel polymer liquid crystal, *Adv. Mater.* **13**, pp. 1337–1339.

Kawatsuki, N., Yamashita, A., Fujii, Y., Kitamura, C., and Yoneda, A. (2008). Thermally enhanced photoinduced reorientation in photo-cross-linkable liquid crystalline polymers comprised of cinnamate and tolane mesogenic groups, *Macromolecules* **41**, pp. 9715–9721.

Kawatsuki, N., Yamashita, A., Kondo, M., Matsumoto, T., Shioda, T., Emoto, A., and Ono, H. (2010). Photoinduced reorientation and polarization holography in photocross-linkable liquid crystalline polymer films with large birefringence, *Polymer* **51**, pp. 2849–2856.

Kempe, C., Rutloh, M., and Stumpe, J. (2003). Photo-orientation of azobenzene side chain polymers parallel or perpendicular to the polarization of red HeNe light, *J. Phys.: Condens. Matter.* **15**, pp. S813–S823.

Okano, K., Shishido, A., and Ikeda, T. (2006). An azotolane liquid-crystalline polymer exhibiting extremely large birefringence and its photoresponsive behavior, *Adv. Mater.* **18**, pp. 523–527.

Okano, K., Tsutsumi, O., Shishido, A., and Ikeda, T. (2006). Azotolane liquid-crystalline polymers: Huge change in birefringence by photoinduced alignment change, *J. Am. Chem. Soc.* **128**, pp. 15368–15369.

Uchida, E., Shiraku, T., Ono, H., and Kawatsuki, N. (2004). Control of thermally enhanced photoinduced reorientation of polymethacrylate films with 4-methoxyazobenzene side groups by irradiating with 365 nm and 633 nm lights and annealing, *Macromolecules* **37**, pp. 5282–5291.

Uchida, E., Kawatsuki, N., Ono, H., and Emoto, A. (2005). Extremely large degree photoinduced in-plane reorientation in azobenzene-containing polymer liquid crystal film using 633 nm light, *Jpn. J. Appl. Phys.* **44**, pp. 570–574.

Uchida, E. and Kawatsuki, N. (2006a). Photoinduced orientation in photoreactive hydrogen-bonding liquid crystalline polymers and liquid crystal alignment on the resultant films, *Macromolecules* **39**, pp. 9357–9364.

Uchida, E. and Kawatsuki, N. (2006b). Influence of wavelength of light on photoinduced orientation of azobenzene-containing polymethacrylate film, *Polym. J.* **38**, pp. 724–731.

Moritsugu, M., Ishikawa, T., Kawata, T., Ogata, T., Kuwahara, Y., and Kurihara, S. (2011). Thermal and photochemical control of molecular orientation of azo-functionalized polymer liquid crystals and application for photo-rewritable paper, *Macromol. Rapid Commun.* **32**, pp. 1546–1550.

Weigert, F. (1919). Dichroism induced in a fine-grain silverchloride emulsion by a beam of linearly polarized light, *Verh. Dtsch. Phys. Ges.* **21**, pp. 479–483.

Wu, Y., Ikeda, T., and Zhang, Q. (1999a). Three-dimensional manipulation of azo polymer liquid crystal by unpolarized light, *Adv. Mater.* **11**, pp. 300–302.

Wu, Y., Mamiya, J., Kanazawa, A. Shiono, T., Ikeda, T., and Zhang, Q. (1999b). Photoinduced alignment of polymer liquid crystals containing azobenzene moieties in the side chain. 6. Biaxiality and three-dimensional reorientation, *Macromolecules* **32**, pp. 8829–8835.

Yu, H. F., Iyoda, T., Okano, K., Shishido, A., and Ikeda, T. (2005). Photoresponsive behavior and photochemical phase transition of amphiphilic diblock liquid-crystalline copolymer, *Mol. Cryst. Liq. Cryst.* **443**, pp. 191–199.

Yu, H. F., Li, J., Ikeda, T., and Iyoda, T. (2006a). Macroscopic parallel nanocylinder array fabrication using a simple rubbing technique, *Adv. Mater.* **18**, pp. 2213–2215.

Yu, H. F., Iyoda, T., and Ikeda, T. (2006b). Photoinduced alignment of nanocylinders by supramolecular cooperative motions, *J. Am. Chem. Soc.* **128**, pp. 11010–11011.

Zebger, I., Rutloh, M., Hoffmann, U., Stumpe, J., Siesler, H. W., and Hvilsted, S. (2003). Photoorientation of a liquid-crystalline polyester with azobenzene side groups: Effects of irradiation with linearly polarized red light after photochemical pretreatment, *Macromolecules* **36**, pp. 9373–9382.

Chapter 4

Low-Molecular-Weight Liquid Crystals

Presently, almost all the liquid crystal (LC) compounds for practical applications in displays are low-molecular-weight (LMW) or small-molecule due to their low viscosity and quick electro-optical response. Incorporating a chromophore with a low-molecular-weight liquid crystal (LMWLC) system, a photoresponsive small molecule LC can be elegantly tailored. In some cases, chromophore groups such as azobenzene and stilbene may act as mesogens, showing an LC phase. However, the most convenient way to prepare photoresponsive LCs is to mix photoinert LMWLCs with a small ratio of photochromic compounds (as impurity). In both of the cases, photoinduced phase transition and photoalignment have been intensively studied. Furthermore, photodriven motion of microscopic objects dispersed in such light-responsive LC systems has also been studied.

4.1 Photoinduced Phase Transition

A photochromic molecule shows reversible photochemical reactions (Chapter 3, Section 3.1), such as photoisomerization (*trans–cis*) and photocrosslinking (e.g., (2+2) photodimerization), which is often accompanied by a change in its molecular configuration. Azobenzenes are one of the typical chromophores, famous for their photoisomerization with a large change in molecular structures, which have been widely used as light-triggers for photoinduced

---

Dancing with Light: Advances in Photofunctional Liquid-Crystalline Materials
Haifeng Yu
Copyright © 2015 Pan Stanford Publishing Pte. Ltd.
ISBN 978-981-4411-11-0 (Hardcover), 978-981-4411-12-7 (eBook)
www.panstanford.com

phase transitions of LMWLCs. As shown in Fig. 3.3, Chapter 3, the two isomerization states are a thermally stable *trans* (also denoted E) and a meta-stable *cis* configuration (the Z state). The longer the excitation wavelength of the actinic light is, the greater the proportion of the *trans* configuration in the photostationary state is. The *cis*-azobenzene isomer can thermally relax back to the *trans*-isomer with a lifetime that strongly correlated with the particular substitution in azobenzene rings.

According to the lifetime of the *cis*-isomer, three kinds of azobenzenes chromophores have been summarized by Natansohn and Kumar [Kumar & Neckers, 1989; Natansohn & Rochon, 2002]. As shown in Fig. 4.1, the first "azobenzene" has relatively poor π–π^* and n–π^* absorbance overlap and the lifetime of the *cis*-isomer is relatively long. The second one is "amino-azobenzene", and there is significant overlap of the two bands and the *cis*-isomer lifetime is shorter. The third azobenzene is "pseudostilbene", where the azobenzene is usually substituted with electron-donor and electron-acceptor substituent.

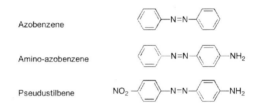

Figure 4.1 Typical examples of three kinds of photoisomerizable azobenzenes classified by Natansohn and Kumar.

4.1.1 Photoinduced Phase Transition in Pure Photochromic LMWLCs

Some azobenzene molecules with substituent of soft spacers can self-assemble into an LC phase (Fig. 3.16 Chapter 3). But the LC phases are observed only in *trans*-azobenzene with a rod-like molecular shape; the *cis*-azobenzene never demonstrates any LC phase at any temperature because of its bent shape. Thus, photoinduced LC-to-isotropic phase transition occurs simultaneously with photoisomerization from *trans*-azobenzenes to their *cis*-isomers upon UV irradiation.

Figure 4.2 gives one example of one azobenzene compound, 4-ethoxy-4′-(6-hydroxy hexyloxy) azobenzene, showing an LC phase in 104–129°C upon cooling. Its colorful four-arm texture was clearly observed with a polarizing optical microscope (POM) at 110°C. At the LC temperature, a dark image is easily induced in the UV-irradiated area due to the loss of birefringence in its isotropic phase (a *cis*-azobenzene rich state). The LC phase can be restored upon irradiation with visible light or thermal treatment because of the completely reversible *cis*-to-*trans* isomerization of the azobenzene LMWLC compound.

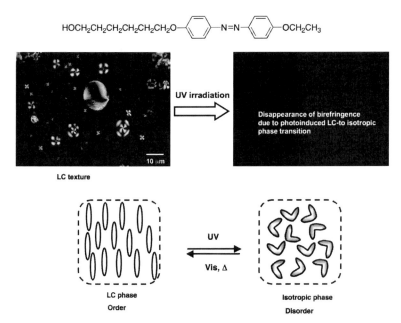

Figure 4.2 Photoinduced phase transition of one azobenzene LMWLC compound.

4.1.2 Photoinduced Phase Transition in Azobenzene-Doped LCs

Even though the azobenzene molecules do not show any LC phase, they can be applied to photomanipulate light-insensitive LCs doped with a low content of azobenzenes via the LC inherent property of molecular cooperative motion (MCM), as shown in Fig. 4.3. The bend shape of the *cis*-azobenzene can destabilize the LC phase self-organized from rod-like molecules. If a small molar proportion of LC molecules change their alignment in response to an external stimulus (like light), the other inactive LC molecules also change their alignment, coinciding with the pre-aligned ordered LCs. In fact, only a small amount of input energy as to cause an alignment change of about 1 mole% of LC molecules is enough to produce the alignment change of the whole LC mixtures. That is, to say, a huge amplification of ordering is possible in LC photonic systems owing to the photoinduced MCM. When a small amount of photochromic molecules (e.g., azobenzene) is incorporated into LC molecules and the resulting host–guest mixtures are irradiated, an LC-to-isotropic phase transition of the mixtures can be isothermally photoinduced. Here, the host represents the majority and gust is the minority in the LC mixture.

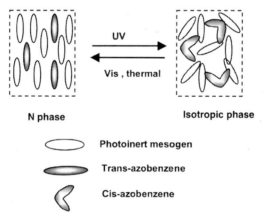

Figure 4.3 Scheme of photochemical phase transition of azobenzene-doped photoinert LC mixture systems by MCM.

The first example of photoinduced LC-to-isotropic phase transition in azobenzene-doped nematic LC mixtures was reported in 1987 [Tazuke et al., 1987]. Since the *trans*-azobenzene is in a rod-like shape, which stabilizes the phase structure of an LC phase, while its *cis*-isomer is bent and tends to destabilize the phase structures of the LC mixture. The LC-to-isotropic phase transition temperature (T_c) of the mixture with the *cis*-form (T_{cc}) is much lower than the mixture with its *trans*-isomer (T_{ct}), as shown in Fig. 4.4. If the temperature of the mixture is placed between T_{ct} and T_{cc}, and the sample is irradiated to induce *trans*-to-*cis* photoisomerization of the azobenzene guest molecules, then T_c decreases with an accumulation of the number of *cis*-isomer. When T_c grows to be lower than the irradiation temperature, the LC-to-isotropic phase transition is thermally induced.

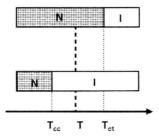

Figure 4.4 Phase diagram of the photochemical phase transition of azobenzene-doped photoinert LC mixture systems (N, nematic LC phase; I, isotropic phase).

Therefore, the photoinduced phase transitions are interpreted in terms of a change in the phase transition temperature of LC host–guest systems on an accumulation of one isomer of the photochromic guest molecule. As expected from Fig. 4.4, ΔT (= T_{ct} – T_{cc}) is one of the most important parameters in the phase transition. When the temperature of LC mixtures is adjusted below T_{cc}, no phase transition can be induced even irradiation with a high dose. When the temperature is regulated close to T_{ct}, the amount of *cis*-azobenzene needed to lower T_c below the irradiation temperature is small, enabling one to effectively photoinduced phase transition.

Photoisomerization of azobenzene molecules is usually reversible. Accompanying with *cis*-to-*trans* back-isomerization, the dye-doped LC sample can revert to the initial LC phase by photo or thermal treatment. This means that phase transitions of LC systems can be reversibly induced by photochemical reactions of photoresponsive guest molecules. Furthermore, the molecular structures of the chromophores greatly influence the photoinduced phase transition of dye-doped LC mixtures. To elucidate the mechanism of photochemical phase transition of the mixture, photoinduced change in azobenzene–4-pentyl-4′-cyanobiphenyl (5CB) systems was in situ evaluated. The LC samples were put between a pair of two-crossed polarizers and the transmittance of probe light through them was measured. If the sample is in an LC phase, non-zero transmittance is observed due to birefringence of the LC properties, while if it is in an isotropic phase, the transmittance should be zero and the dark spot should be observed with a polarizing optical microscope (POM). As shown in Fig. 4.5, upon photoirradiation of 5CB/*trans*-azobenzene, dark spots in the light-irradiated areas of the samples are observed with a POM, due to the formation of isotropic domains, where a nematic LC-to-isotropic phase transition is induced and leads to local phase-segregated domains between the LC phase and the isotropic phase. With the increase in the irradiated dose, the isotropic domains grow and the dark areas increase and finally, the whole vision area becomes dark and isotropic due to the occurrence of photoinduced phase transition [Sung et al., 2002].

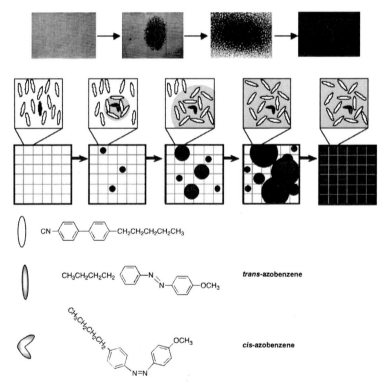

Figure 4.5 Experimental results and their reasonable schemes of photochemical phase transition in azobenzene-doped photoinert LMWLC (5CB) systems.

The molecular shape of each isomer of photochromic molecules shows great influence on the pohotochemical phase transitions of LMWLCs. As described above, 4,4′-disubstituted azobenzenes possess a rod-like shape in the *trans* form while the *cis*-isomers show a bent shape (Fig. 4.6). This difference in the molecular shape enables its photoisomerization to produce a significant change in the stabilization of phase structures of LCs. If both isomers possess a similar molecular shape, the photoinduced effect will be little. A good example is 3,3′-disubstituted azobenzenes (Fig. 4.6). In this azobenzene derivative, both *trans*- and *cis*-isomers show a rod-like molecular structure, and *trans*–*cis* isomerization only induces little change in the molecular shape. As a result, photoirradiation of a mixture of 3,3′-disubstituted azobenzene and a photoinert nematic LC caused no photochemical phase transition even in the presence of as high as 20 wt.% of the 3,3′-disubstituted azobenzene molecule.

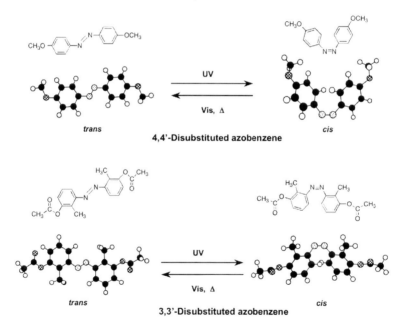

Figure 4.6   Photoisomerization of 4,4′-disubstituted and 3,3′-disubstituted azobenzene derivatives. Upon UV irradiation, the former shows a significant change in the molecular shape, and the latter exhibits little change in the molecular shape.

The photochemical phase transition in the azobenzene-doped photoinert LC mixture has been systematically studied for a series of LC molecules. When the photochromic molecules alter their molecular shapes upon photoirradiation, modifications in other properties such as polarity are usually generated. Such photochromic reactions are reversible and two isomers can be effectively interchanged by light with different wavelengths. Furthermore, most of photochromic reactions are very fast and occur in a time region of pico seconds (ps).

4.1.3 Photoinduced Phase Transition in LCs Doped with other Dyes

Apart from azobenzenes, various photochromic molecules such as stilbenes, spiropyrans, diarylethenes, and fulgides have been adopted to study the photoinduced phase transition [Kurihara et al., 1991]. Kurihara et al. investigated photochemical phase transition behavior of mixtures of spiropyran derivatives and nematic LCs, and found that merocyanine (open-ring form) rather stabilizes phase structures of nematic LCs due to its linear structure, whereas spiropyran (closed-ring form) destabilizes the phase structures (Fig. 4.7e). It was observed that photoirradiation of a mixture of merocyanine and a nematic LC with visible light to cause merocyanine–spiropyran isomerization induces a nematic-to-isotropic phase transition and upon irradiation of a mixture of spiropyran and a nematic LC with UV light the initial nematic phase is restored.

Fulgides are extensively studied photochromic compounds (Fig. 4.7d) [Allinson & Gleeson, 1993; 1995]. Gibbons et al. [1991] investigated the effect of photoisomerization of furylfulgide dispersed in a cyanobiphenylnematic LC (E7) at a concentration of lower than 2 wt.% on the T_c of the mixture. It was found that ΔT observed upon photoirradiation is small. They also studied the effect of photoisomerization on the physical properties of the nematic LC (dielectric constants and elasticity) and ascribed the observed change in these properties to the change in T_c upon photoisomerization.

Diarylethenes are a new class of photochromic molecules that show high thermal stability of both isomers and high fatigue resistance (high durability). A diarylethene with a chiral moiety

(Fig. 4.7b) was found to act as a chiral dopant that can induce a cholesteric phase (chiral nematic) [Denekamp & Feringa, 1998]. The diarylethene compound was added to a cyanobiphenyl nematic LC at a concentration of 0.7 wt.% to induce a cholesteric phase. The cholesteric mixture was then irradiated with UV light at 300 nm, which caused disappearance of the chiral nematic phase and appearance of a nematic phase. Upon irradiation of the nematic phase with visible light, the initial cholesteric phase reverted. This reversible phase transition was interpreted in terms of the helical twisting power (HTP) arising from the molecular shape of the photoactive dopant: the closed-ring isomer exhibits a smaller HTP than the open-ring isomer. However, this is not always the case. The HTP sensitively depends on the structure of the arylethene. In another compound, the closed-ring isomer shows a larger HTP than the open-ring form [Yamaguchi, 2000]. When a diarylethene unit was incorporated into a chiral cyclohexane, the resultant molecule showed a larger HTP in the closed-ring form. The cyclohexane with two diarylethene moieties was dispersed in a cyanobiphenyl nematic LC and the resulting mixture was exposed to UV light to cause an isomerization from the open-ring form to the closed-ring form. Different from azobenzene molecules, this isomerization also induced a nematic-to-cholesteric phase transition.

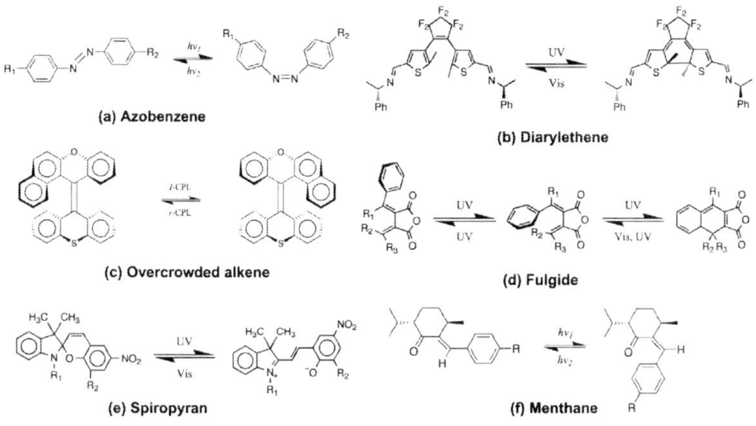

Figure 4.7 Various photochromic molecules used in the photochemical induced change, (a) azobenzene; (b) diarylethene; (c) overcrowded alkenes; (d) fulgide; (e) spiropyran; (f) menthane. (UV: ultraviolet light; Vis: visible light; r-CPL: right circularly polarized light; l-CPL: left circularly polarized light.)

4.1.4 Photoinduced Phase Transition in LMWLCs–Polymer Composites

Generally, microphase separation occurring in LMWLCs–polymer composites scatters the visible light because of the formed large domain size (in an order of microns), resulting in poor optical transparence (an opaque state). Exerted with an external electric field, a highly transparent sate can be obtained by eliminating the scattering from the micron-sized domains. The advantages of the composite film are that they do not need any polarizers to show the scattering–transparent contrast and avoid the additional treatment for LC alignment, which is far different from LC cells for display applications, and enables them to exhibit high processability and flexibility.

Recently, they have been extensively explored for practical applications for displays. Several kinds of LMWLCs–polymer composites have been explored, including nematic curvilinear aligned phase materials with an encapsulated LC structure, polymer-dispersed LCs (PDLCs) with LC droplets dispersed in a polymer matrix by means of polymerization-induced or solvent-induced phase separation, polymer network LCs with micron-scaled LC domains, and polymer-stabilized LCs with a small amount of polymer network. Similarly, photoinduced phase transition in the microphase-separated domains of LMWLCs–polymer composites may give rise to novel optical effects.

Takizawa et al. [1994] reported a photonic application of the LMWLCs–polymer composite films: a spatial light modulator (SLM) that displays large images on a screen by projecting an image created on a small valve (Fig. 4.8). The PDLC-SLM mainly consists of two components: a PDLC layer, which modulates the probe light and a photoconductive layer ($Bi_{12}SiO_{20}$), which detects the stimulus light. In the initial state, some voltage is applied as a bias to the two layers. Upon irradiation with the stimulus light, the resistance of the photoconductive layer decreases. The voltage, therefore, becomes applied mainly to the PDLC layer. The increase in the voltage across the PDLC layer brings about unidirectional alignment of LC droplets. Consequently, the probe light reflected with the dielectric mirror can be modulated due to transformation between the light-scattering state and the transparent state in the PDLC layer. A high contrast ratio of 178:1, a response time of 14 ms, and a decay time of 15 ms have been achieved.

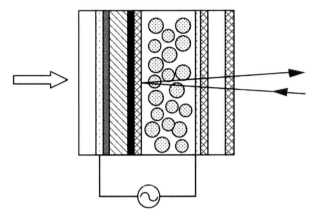

Figure 4.8 Spatial light modulator (SLM) with a PDLC structure.

It was reported that light addressing and optical image recording by means of LMWLCs–polymer composite films showing electric field frequency-dependent optical properties [Kajiyama et al., 1993]. As shown in Fig. 4.9, a composite film consisting of a side-chain LCP with a polysiloxane backbone and LMWLCs exhibits a remarkable light scattering property upon application of a low-frequency electric field, while it is highly transparent upon application of a high-frequency electric field. An azobenzene derivative was added to the composite film as a photoresponsive molecule. The threshold frequency, defined as the critical frequency at which the composite films change from a transparent state to a turbid one, was found to change upon irradiation. The threshold frequency of the composites containing *trans*-azobenzene, f_c, was smaller than that of the composites with *cis*-azobenzene, f_{cs} (Fig. 4.9). The transmittance of the composites can be switched by stimulation in the presence of an electric field with frequency of f_d, which is between f_c and f_{cs} as shown in Fig. 4.9. Since both transparent and turbid states of the composite films are stably memorized even after removal of electric fields, they can be applied to rewritable optical image recording media. Laser irradiation has been used for the formation of positive and negative images through photothermal processes.

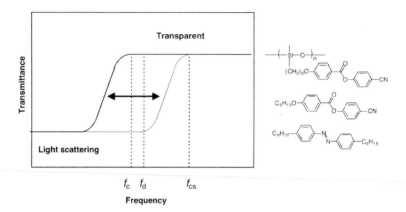

Figure 4.9 Change in the optical properties of polymer–LC composite film upon photoirradiation as a function of frequency.

In confined polymer matrices with microphase separation, the photoresponsive LMWLCs–polymer composite films driven by photon-mode process without electric fields have been achieved via photochemical phase transitions. LMWLCs–polymer composite films with a thickness of 2–3 μm using a mixture of a nematic LC and an azobenzene derivative dispersed in an aqueous solution of polyvinyl alcohol (PVA) by solvent-induced phase separation, showed very low transmittance because of opacity. But they became transparent upon irradiation at 366 nm, resulting from nematic LC to isotropic phase transitions of LC droplets within the polymer matrix due to *trans–cis* photoisomerization of azobenzenes (Fig. 4.10a). The recovery of the initial opaque state could be achieved by irradiation with visible light to cause *cis–trans* back-isomerization of the azobenzenes. In this system, the degree of change in transmittance was as low as 10–50%.

To improve the optical properties of the LMWLCs–polymer composite films, the polymerization of a ternary mixture of bifunctional acrylate monomers, nematic LCs, and azobenzene compounds was carried out in a cell with a 10 μm gap and their optical properties were evaluated [Lee et al., 1998]. It was found that the transmittance of these composite films could be photomodulated in a range approximately from 0 to 100% by photochemical phase transitions (Fig. 4.10b). Furthermore, optical image storage and reverse-mode switching were achieved in such azobenzene-containing polymer network systems by choosing network-forming materials and by tuning polymerization conditions.

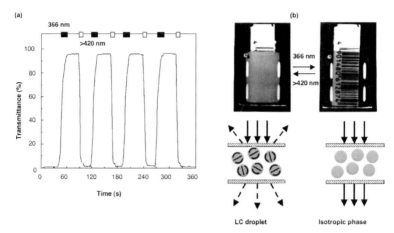

Figure 4.10 Photochemical phase transition in polymer–LMWLC composite films. (a) Change in transmittance of polymer–LMWLC composite films upon irradiation. (b) Photos of sample films in the initial opaque state became highly transparent after UV irradiation due to the phase transition of the LC droplets.

4.2 Phototunning of Cholesteric LCs

4.2.1 General Principles

Cholesteric LCs (CLCs) can self-assemble into a helically ordered structure with a helical pitch (p) in an order of wavelength (λ) of visible light. One of their unique optical properties is selective reflection of light when their helix axes are perpendicular to the surface of LC cells. The wavelength of the reflective light is determined by the helical pitch of CLCs (Fig. 4.11), which can be calculated by $\lambda = np$ (n, the mean refractive index of CLCs).

If the helical pitch of CLCs can be phototuned by light with a suitable wavelength, photocontrol of selective reflection could be achieved accordingly (Fig. 4.12). Furthermore, since the helical pitch is very sensitive to external stimuli such as temperature, pressure, and impurity, the wavelength of the selective reflection is also strongly dependent on these factors. Besides, CLCs are expected as active media for reflection-type displays, particularly if the wavelength of the reflective light is in the visible region, these CLC materials are highly potential for full-color LCDs with high quality and extremely low energy consumption [Hu et al., 2010].

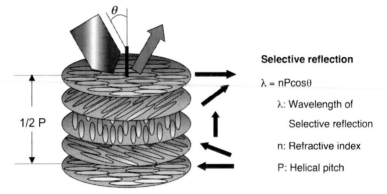

Figure 4.11 Scheme of selective reflection of CLCs.

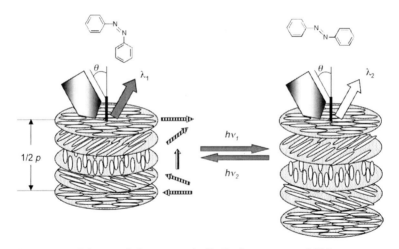

Figure 4.12 Scheme of photocontrol of helical structures of CLCs.

4.2.2 Phototuning CLCs with Photoisomerization

It has been demonstrated that CLCs can be induced in a nematic LC system by doping chiral compounds (Fig. 4.13). This provides a good chance for phototuning of helical structures via introducing photochromic molecules. HTP, defined as the capability of a chiral group to induce cholesteric LC phase in a nematic LC host, can quantitatively describe the photocontrol of cholesteric LCs. HTP = $1/pc$, where c is the dopant concentration. Thus, a higher HTP value requires less chiral dopant to yield the same value of wavelength. HTP depends on the dipole–quadruple interactions of the chiral molecules with its nematic neighbors, the anisotropy of the nematic host phase, and the order parameter. Sackman [1971] first reported the photochemical change in alignment of cholesteric LCs by irradiating a mixture of nonchiral azobenzene-dispersed CLCs. It was found that the pitch could be altered by photoisomerization of azobenzenes, which was confirmed by the change in the wavelength of reflection light (Fig. 4.12).

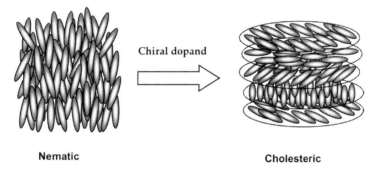

Figure 4.13 Preparation of CLCs by doping nematic LCs with chiral compounds.

In a chiral nematic LC system doping with one achiral azobenzene compound, the helical pitch can be photoruned upon UV photoirradiation. Figure 4.14 shows the mechanism of the photoinduced change in selective reflection by the photoisomerization. In such an LC system, a higher concentration of *trans*-azobenzene and a higher temperature can induce a lower helical pitch.

The optical switching of selectively reflective colors was investigated based on photoinduced change in the helical pitch of CLCs doped with a chiral azobenzene [Lee et al., 2000]. Doping of cyanobiphenyl nematic LCs with chiral dopants possessing strong HTP produced CLC phases with reflection bands in a visible region. Photoirradiation of the CLCs containing the chiral azobenzene resulted in a change in the helical pitch due to *trans–cis* isomerization of the chiral azobenzene, followed by the change in the reflective wavelength. Such a change in the helical pitch can be interpreted in terms of the change in the phase transition temperature caused by changing the molecular shape of the guest azobenzenes.

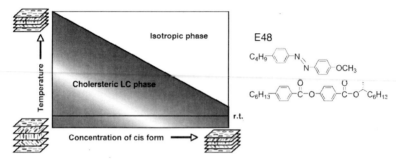

Figure 4.14 Mechanism of photoinduced change in selective reflection of CLCs.

White et al. [2010] demonstrated the reflection bandwidth of CLCs consisting of a chiral azobenzene molecule possessing strong helical twisting-power could be broadened from 100 nm to as much as 1700 nm. The phototuning of more than 2000 nm could be achieved for the CLCs composed of nematic LCs and the chiral azobenzene compound [White et al., 2009].

4.2.3 Phototuning with other Chromophores

In addition to azobenzene chromophores, many studies have been carried out on phototuning of helical structures and properties of CLCs through photoisomerization of various types of photochromic compounds, with or without chiral groups in molecules, such as stilbenes, diarylethenes, overcrowded alkenes, menthones, flugides, spiropyran, and so on. In these CLC systems, the photochromic compounds with chiral groups are dissolved in nematic LCs, and the induced helical structures are photocontrollable, leading to a difference in HTP among isomers through photochemically induced change in molecular configurations.

Recently, Tamaoki [2001] developed a series of dicholecteric ester oligomers, which can fix the reflection color by immobilizing the molecular ordering in a glassy state and shows applicable to optics materials with a static function. Eelkema & Feringa [2006] reported fully reversible control of the reflection color of this film across the entire visible spectrum by using a chiral molecular motor as a dopant in a CLC film. As shown in Fig. 4.15, a large difference in HTP between the two isomeric forms of the motor allows efficient light and thermal-induced switching of the helicity of the CLC superstructure.

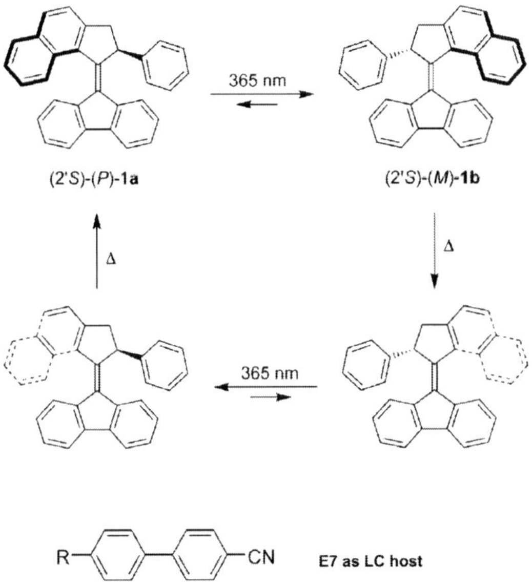

Figure 4.15 Photoreaction of one molecular motor for phototuning CLCs by doping is in LMWLC (E7). Reprinted with permission from Eelkema and Feringa, Copyright 2006, WILEY-VCH Verlag GmbH & Co. KGaA, Weinheim.

4.2.4 Possible Applications

In the photoresponsive CLC systems, the unique properties of reflecting light in controllable wavelength are based on their helical supramolecular conformations. Modification of their orientation states may change the input light information, thus three states of reflection, scattering, and transparence can be obtained accordingly when CLCs are in planar, random, and vertical alignment, respectively. As shown in Fig. 4.16, CLCs in planar alignment exhibit a planar texture under a POM, and they reflect a certain wavelength light depending on their helical pitch. This can be used as 1D photonic band-gap materials due to the periodical helical structures. Without the function of any external fields, CLCs are in random alignment, and optical scattering often happens. At this case, focal conic texture can be observed [Yang et al., 1997]. When an electric field is applied, the helical structure of CLCs can be unwound, leading to a vertical alignment. Light can permeate them completely.

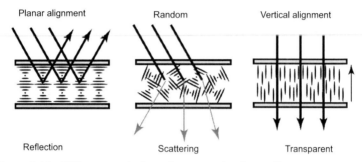

Figure 4.16 Different optical performance in three alignment states of CLCs.

In the planar alignment state of the photoresponsive CLCs, the reflective light can be photocontrolled by the isomerization of azobenzenes. Figure 4.17 shows phototuning of reflection from blue to red through green in CLCs doped with chiral azobenzenes [Li et al., 2011]. All the reflection color images were obtained from LMWLC host E7 doped with 6.0 wt.% chiral azobenzene in a 5 µm thick planar LC cell. These might provide a good method to realize full-color LCD in reflection mode.

Photoirradiation can also reversibly control a nematic to CLC phase transition. In a mixture of LC E44 doped with one chiral photo-inert compound and one chiral azobenzene compound shown in Fig. 4.18, a compensated nematic phase was changed into CLC phase on UV irradiation [Alam et al., 2007]. A schlieren texture in the initial state (the below left-hand of Fig. 4.18) indicates that the LC exists in the compensated nematic state. The compensated nematic state was destroyed upon UV irradiation, because of the photochemical isomerization of the chiral azobenzenes. In addition, Fig. 4.18 indicates that the helical pitch was decreased by UV irradiation time. Reversible photoisomerization from the *cis* form to the *trans* form was observed on visible light irradiation, and HTP was compensated once more, so the compensated nematic phase was recovered again upon visible light irradiation.

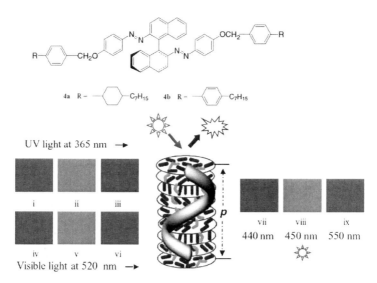

Figure 4.17 Photocontrol of various VIS spectrum wavelengths' reflection of CLCs doped with chiral azobenzenes. Reprinted with permission from Li et al., Copyright 2011, WILEY-VCH Verlag GmbH & Co. KGaA, Weinheim.

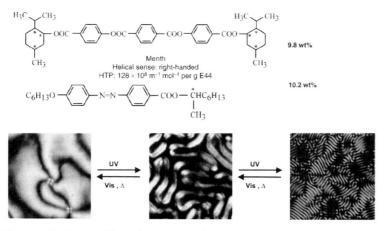

Figure 4.18 Reversible photochemical phase transition between a compensated nematic phase and a CLC phase by UV and visible light irradiation. Reprinted with permission from Alam et al., Copyright 2007, WILEY-VCH Verlag GmbH & Co. KGaA, Weinheim.

4.3 Photochemical Flip of Polarization of Ferroelectric LCs

As shown in Fig. 4.19, ferroelectric LCs (FLCs), with an LC phase structure of chiral smectic C (SmC*), possess spontaneous polarization (*Ps*) and exhibit microsecond responses to the change in applied electric fields in a surface-stabilized state (flip of polarization). If a flip of polarization of surface-stabilized FLCs (SSFLCs) can be photoinduced in the presence of an applied electric field, a photoresponse of FLCs in the microsecond time region can be obtained.

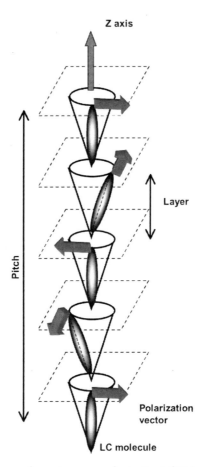

Figure 4.19 Scheme of spontaneous polarization of FLCs.

A FLC mixed with 3 mole% azobenzene was subjected to the surface-stabilized state in a very thin LC cell (thinner than the pitch), which was then irradiated at 366 nm to cause *trans–cis* photoisomerization of the guest azobenzenes. It was found that a threshold electric field for the flip of polarization (coercive force) of SSFLCs was changed upon photoirradiation [Ikeda et al., 1993]. SSFLCs show a hysteresis between the applied electric field and the polarization in Fig. 4.20. It was observed that the hysteresis of the mixture of *trans*-azobenzene and FLC was different from that of the mixture one of *cis*-azobenzene and FLCs. This influence of molecular shape on the coercive force is very similar to the different T_c observed in the azobenzene/LMWLC mixtures. In the mixture of azobenzene and FLCs, when the azobenzene is in the *trans* form, it no longer disorganizes the SmC* phase structure of the SSFLCs significantly. Whereas, the phase structure of the SmC* LC is seriously affected when the azobenzene is in the *cis* form, resulting in much reduction of the threshold value for the flip of polarization.

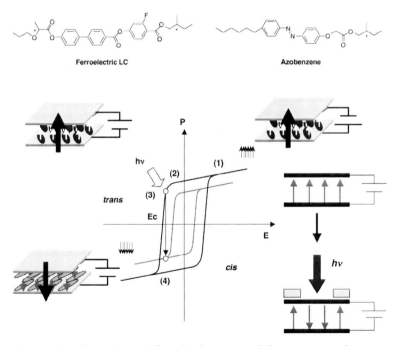

Figure 4.20 Photochemical flip of polarization of the mixture azobenzene and FLCs.

Based on these properties shown in Fig. 4.20, a new mode of optical switching of FLCs has been proposed [Sasaki et al., 1994]. The working principle can be described: (1) Polarization of a SSFLC cell containing a small amount of azobenzene molecules is aligned along one direction by an electric field; (2) An opposite electric field is applied across the SSFLC cell, which is small enough to keep the initial direction of the polarization unchangeable; (3) With this field as a bias, the SSFLC cell is irradiated to cause photoisomerization of the azobenzene, then the change in hysteresis of the cell is induced; (4) The threshold value for the flip of polarization is lowered upon photoirradiation and becomes smaller than the bias voltage, which certainly induces the flip of polarization of SSFLCs. In other words, the bias voltage remains unchangeable before and after photoirradiation, but the threshold value for the flip of polarization is reduced by *trans–cis* photoisomerization of the azobenzenes. As a result, a flip of polarization is induced upon photoirradiation, which leads to a change in alignment of the FLCs.

SSFLCs show bistable polarization with upward and downward directions with respect to normal to the cell surface, and hence both of the two alignment states of SSFLCs are stabilized. Furthermore, these two states remain unchanged even after the electric field is removed. Owing to these properties of SSFLCs, once the flip of polarization is induced upon photoirradiation, the direction of the polarization is opposite between the irradiated and unirradiated sites, and the alignment of FLCs is different between the two sites. These changes in polarization and alignment of SSFLCs engender an optical contrast between the irradiated and unirradiated sites, and they remain unchanged (memory effect).

Time-resolved measurements of the change in alignment due to the flip of polarization in the mixtures of azobenzene and FLCs were performed upon pulse irradiation with the third harmonic of a yttrium aluminium garnet (YAG) laser (355 nm; fwhm, 10 ns) and a flip of polarization in 500 µs was observed [Sasaki et al., 1994], which is the first example to photochemically induce a flip of polarization of SSFLCs (Fig. 4.21). Then detailed studies have been performed on the flip of polarization in the photochromic guest/FLC host systems: effects of the structure of FLC hosts, the structure of photochromic guests, temperature, bias voltage, and the change in *Ps*.

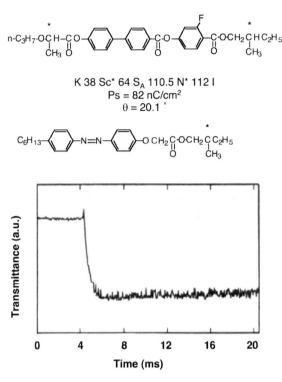

Figure 4.21 Time-resolved observation of photochemical flip of polarization in a mixture of azobenzene and FLCs. Reprinted with permission from Sasaki et al., Copyright 1994, American Chemical Society.

In Fig. 4.22a, antiferroelectric LCs (anti-FLCs) showed a similar behavior of photochemical flip of polarization to the azobenzene/anti-FLC mixtures, which was examined upon irradiation to cause *trans–cis* photoisomerization of the guest molecule [Moriyama et al., 1993]. Some novel guest chromophores might effectively induce photochemical flip of polarization in FLC systems. An azobenzene derivative with a chiral cyclic carbonate was designed on the basis of a large value of polarization, resulting from a chiral cyclic carbonate structure, and examined as a chiral dopant to induce the SmC* phase (Fig. 4.22b). In this system, the chiral dopant also acts as one photoresponsive molecule, thereby it is expected that a change in molecular shape of the dopant would greatly influence the phase structure of the SmC* LC because the molecular shape of the dopant is crucial for the induction of the SmC* phase. It was found that the azobenzene dopant shown in Fig. 4.22b is quite effective to induce the photochemical flip of polarization of the FLC mixture. Moreover, the photochemical flip of polarization in the anti-FLC is very effectively induced and a device fabricated using these anti-FLCs were explored [Shirota & Yamaguchi, 1997].

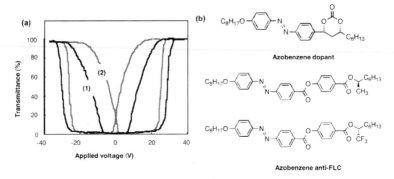

Figure 4.22 (a) Change in transmittance in anti-FLC/azobenzene, (1) in the dark; (2) under photoirradiation. (b) Molecular structures of photoactive chiral azobenzene dopant and azobenzene anti-FLCs.

4.4 Phototriggered Sol–Gel Transition in LMWLC Organogels

Physical organogels based on fibrous self-assembly in organic solvents are one of the powerful tools to prepare functional molecular materials. Apart from hydrogen bonding, lipophilic function and π–π interaction, molecular chirality also plays an important role in the process of molecular self-assembly. For instance, a chiral LMWLC azobenzene dopant with a *syn*-chiral carbonate moiety ((*R,R*)-8AC6 in Fig. 4.23) can be used as an organogelator for various organic solvents, whereas no organogel formed for 8AC6 without chirality in the same solvent [Mamiya et al., 2002]. The physical organogels were constructed by (*R,R*)-8AC6 having a simple structure without hydrogen bonds of amide or urea moieties or stacking of cholesterol groups.

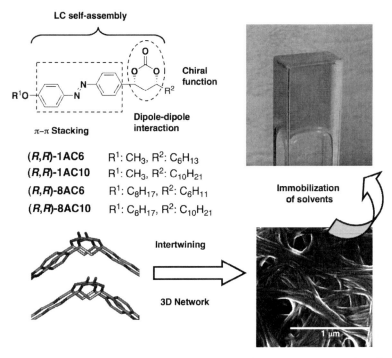

Figure 4.23 Photoinduced sol–gel transition of LMW LC organogel with chiral azobenzene derivatives. The photo and SEM pictures were obtained for physical organogel formed by 8AC6/2-propanol gel (6 g/L) at room temperature.

Various measurements were performed to clarify the self-assembled process of the LMWLC organogels [Mamiya et al., 2002]. As shown in Fig. 4.23, SEM observation of the organogels revealed that (R,R)-8AC6 forms fibrous aggregates in organic solvents and the gelation is due to the formation of 1D fibrous structures. Moreover, the solvent molecules have been incorporated in the periphery of the self-assembled structures of the materials in an LC state, leading to the swollen fibrous aggregates. To evaluate the relationships between the self-assembled properties and the LC phase, the azobenzene derivatives having different lengths of the alkyl chain were synthesized. It was found that the length of alkyl chain in the azobenzene compounds significantly influences crystallinity and solvophobicity as well as the gelation ability. The obtained azobenzene derivatives possessing non-LC properties do not show gelation ability, indicating the LC property plays a critical role in the formation of physical organogels. In addition, the chirality of the carbonate moiety may act as a determining role in the fiber formation and the induction of the gelation. Single-crystal X-ray analysis of an analogous compound of gelator (R,R)-8AC6 was carried out to examine the self-assembled structure. It was demonstrated that the dipole–dipole interaction of the chiral carbonate moiety having a large dipole moment serves as a key role in formation of the ribbon-like structure, leading to the fibrous self-assembly.

The present molecular design of organogelators also leads to the development of novel molecular assemblies with photocontrollable function. Since (R,R)-8AC6 in a non-LC phase did not show gelation capability, the phototriggered sol–gel transition of LC organogels can be easily induced by the LC-isotropic phase transition upon UV irradiation. As shown in Fig. 4.24, the domain size of the networks formed by the fibrous aggregates was similar to the wavelength of visible region, leading to highly light-scattering performance. A completely transparent solution can be induced by the phototriggered sol–gel phase transition of an organogel. In the *cis*-azobenzene form, the formation of 1D fibrous aggregation is prevented due to its bend molecular configuration. The theoretical calculation of (R,R)-8AC6 indicated that photoisomerization induced a change in both molecular shape and polarity. Unlike the *trans* form of (R,R)-8AC6, its *cis* form is soluble in organic solvents because of an increase in the molecular polarity. The photocontrollable process between the transparent solution and the opaque light-scattering organogel can be reversibly caused by *trans–cis* photoisomerization of the azobenzene moiety.

Similar to the LMWLC organogelators, some azobenzene-containing LMW non-LC compounds with specific intermolecular structures can act as gelators in photoinert LMWLC solvents, as shown in Fig. 4.25 [Zhao & Tong, 2003]. Only a small amount of the azobenzene-containing LMW compounds is sufficient to induce formation of LC gels by self-assembly, in which photochemical phase transition was correspondingly achieved. Zhao et al. recorded holographic structures in an LC gel with gelator in Fig. 4.25a and studied their electrically controlled diffraction behaviors [Tong &

Zhao, 2003; Zhao & Tong, 2003]. Moriyama et al. [2003] studied anisotropic gels obtained with a gelator (Fig. 4.25b) in LMWLC 5CB, and achieved the photoinduced reversible structural changes between nematic LCs (in *trans*-azobenzene) and CLCs (in *cis*-azobenzene) at room temperature. Such achievements indicate the potential of exploiting photoresponsive LMWLCs for new soft materials and novel applications.

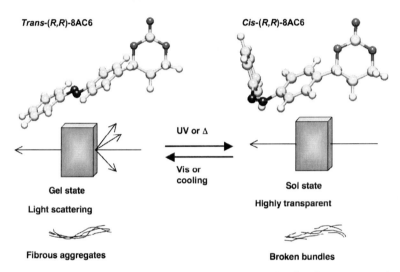

Figure 4.24 Schematic illustration of photoinduced sol–gel transition of LMWLC organogel. The molecular configurations of *trans*- and *cis*-8AC6 in energy minimization were calculated by MOPAC PM3 method.

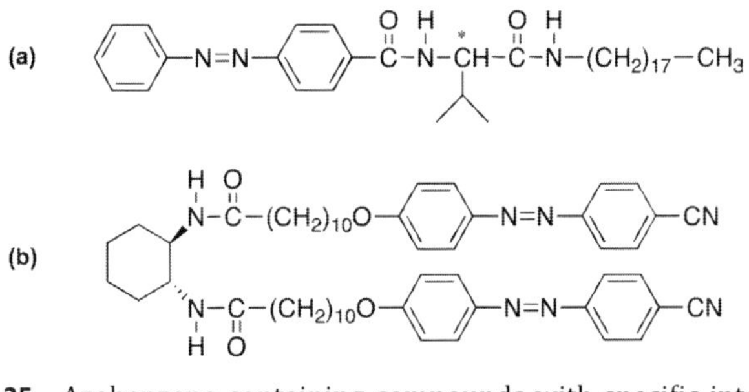

Figure 4.25 Azobenzene-containing compounds with specific intermolecular structures for gelators in LMWLC as solvents (a) for holographic recording (b) for fabrication of anisotropic gel and photoinduced reversible change in molecular structure.

4.5 Photocontrolled Orientation by Photophysical Processes

LCs are highly birefringent phases, so that one can obtain a large change in refractive index by changing the LC alignment other than conventional nonlinear polarization observed in non-LC materials. For example, electric fields can change the orientation of LC directors, which gives rise to a large nonlinear optical (NLO) effect and turns out to be 10^6–10^{10} orders of magnitude greater than the nonlinearity of usual NLO materials [Durbin et al., 1981]. Although the study of NLO effects in LCs was started as early as in 1970s, it is only after the finding of optical alignment changes of transparent LCs by Durbin et al. [1981] that yields a rapid growth in this field. The photoalignment change of LC molecules in transparent LC systems is usually called "optical Fréedericksz transition (OFT)" by analogy of the electric Fréedericksz transition (Fig. 4.26). OFT describes optical field-induced director reorientation: if a continuous linearly polarized laser beam is normally incident on a homeotropically aligned nematic LC cell, the LC alignment would change from a homeotropic state to a homogeneous state when the light intensity is increased beyond a threshold value. This process gives rise to large NLO effects, making LCs the largest nonlinear material. The optical field-induced director reorientation can be interpreted as a result of optical dielectric anisotropy of LCs and a tendency of LC directors becoming aligned parallel to the light electric field to minimize the free energy of the system. The reorientation is driven by an optical field–induced reorientation torque, which induces a rotational motion of the LC director from its original direction to the light polarization direction. At an equilibrium state, the optical reorientation torque is balanced by an elastic torque that originates from the bulk distortion of LCs, with LC molecules being aligned in the polarization plane of incident actinic light.

The optically induced reorientation behavior has been found to strongly depend on light polarization and experiment configurations. For example, if one irradiates a homeotropic LC cell with polarized light perpendicular to its director, the threshold intensity for OFT increases considerably with an increase in an incident angle because deformed LC molecules change the polarization direction of the incident light, while for actinic light polarized parallel to the director, light intensity required for reorientation shows a contrary variation tendency with the incident angle.

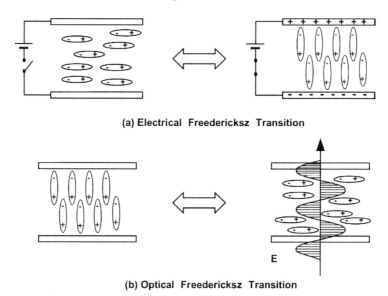

(a) Electrical Freedericksz Transition

(b) Optical Freedericksz Transition

Figure 4.26 Schematic illustration of Fréedericksz transition induced (a) by an electric field and (b) by an optical field.

As an NLO effect, the optically induced director reorientation, which gives rise to a large change in refractive index, can in turn affect the optical properties of the incident beams. The phenomena such as, self-focusing and self-phase modulation have been observed with nematic LC materials. Especially, when a laser beam with a Gaussian traversing intensity distribution is used as an excitation source, the non-uniform director reorientation corresponding to the traversing intensity profile leads to a spatial self-phase modulation. Once the phase difference of the light between the beam center and edge is larger than 2π, self-diffraction of the incident light can be observed.

Upon linearly polarized light (LPL) irradiation, photoresponsive LMWLCs can be photomanipulated into ordered states by photochemical processes. According to the Weigert effect, the transition moments of LMWLC molecules can be photocontrolled perpendicularly to the polarization direction of the actinic light. Azobenzene-containing LCs and azobenzene/LMWLC guest–host systems are typical for such kind of photoalignment. Different from the photochemical processes, Jánossy et al. [1990] found that a small amount of certain anthraquinone dye dissolved in a nematic LC, which makes the system absorb light and undergoes only photophysical process upon excitation. As shown in Fig. 4.27, Jánossy effect induces an LC alignment parallel to the direction of LPL, which has attracted considerable attention in the past decade since such highly light-sensitive dyes are of great interest for optical applications.

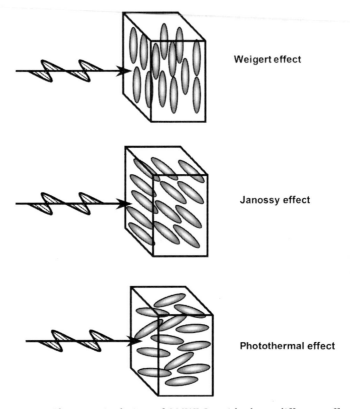

Figure 4.27 Photomanipulation of LMWLCs with three different effects; the Weigert, Jánossy, and photothermal effects induce the alignment of LCs, perpendicular, parallel, and random to the polarization direction of the actinic light.

Similarly, the photoinduced director reorientation of LMWLCs by photophysical processes gives rise to a large change in refractive index. As shown in Fig. 4.28, the dyes can be first photoinduced into an ordered state. Then, the photoinert LC molecules are aligned along the alignment direction of the oriented dyes due to the MCM.

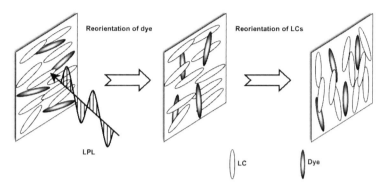

Figure 4.28 Photoinduced reorientation of LMWLCs by photophysical process.

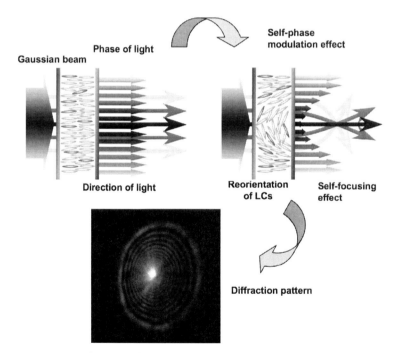

Figure 4.29 Schematic illustration of self-phase modulation effect and self-focusing effect, leading to typical diffraction rings formed on the screen.

On the other hand, the reorientated LCs can in turn affect the optical properties of the incident beams. For instance, interesting phenomena such as self-focusing, self-phase modulation have been observed with nematic LMWLCs [Durbin et al., 1981], as shown in Fig. 4.29. Especially, when a laser beam with a Gaussian traversing intensity distribution is used as an excitation source, the non-uniform director reorientation corresponding to the traversing intensity profile leads to a spatial self-phase modulation. Once the phase difference of the light between the beam center and edge is larger than 2π, self-diffraction of the incident light can be observed, leading to the formation of typical diffraction rings on the screen (Fig. 4.29). This photophysically induced reorientation of LCs is far different from the photochemical process, which is summarized in Table 4.1.

Table 4.1 Comparison of photochemical and photophysical processes in LC reorientation

Ways of processes	Photochemical	Photophysical
Typical materials	(structures shown)	(structures shown)
Molecular weight	LMW and polymers	LMW
Driver force for LC alignment	Photoisomerization photocrosslinking	Intramolecular reciprocity
Principles of LC alignment	Order–disorder–order or disorder–order	Order–order

Various dyes have been used as effective dopants for the photoinduced reorientation by the photophysical processes, as long as they can be dissolved in the LMWLC host and absorb the incident laser beam. Figure 4.30 shows typical dyes used in recent studies of the dye-induced reorientation of LC molecules by the Jánossy effect. The compounds anthraquinone-1 (AQ1) and AQ2 are effective dyes known so far, and they can decrease the threshold light intensity by about 100 times at the concentration of about 1 wt.%. The compounds AQ3 and AQ4 show the same effect but the efficiency is lower than that of AQ1 and AQ2. Kreuzer et al. [2002] investigated the effect of deuterium–hydrogen isotopic substitution of AQ1 on the efficiency of photoinduced reorientation. In accordance with the predictions of their model, the results were well correlated with the increase in a characteristic decay time of the electronically excited dye.

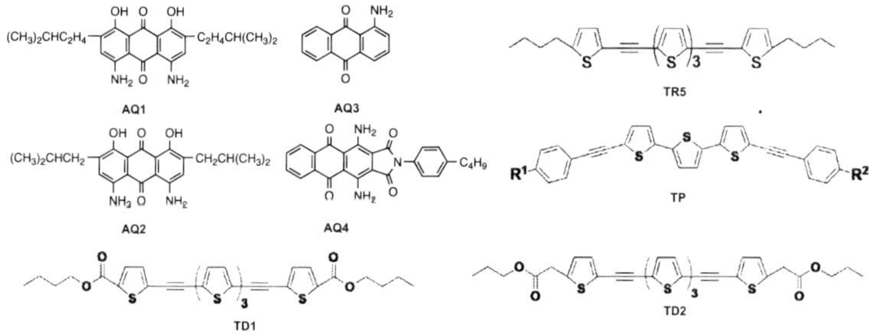

Figure 4.30 Examples of dyes used for photophysically induced orientation of LMWLCs.

Recently, an oligomer of thiophene derivative with linear π-conjugation (TR5 in Fig. 4.30), acts as a highly efficient dye for photoinduced reorientation was reported [Zhang et al., 2000]. It is assumed that the high efficiency of TR5 is mainly due to the change in the molecular polarizability between the ground and excited states, which is related to the intramolecular delocalization of π-electrons along the molecular long axis. However, molecular interactions responsible for the photoinduced reorientation are still unclear. Moreover, the light intensity required for reorientation of LCs is still as large as several W/cm^2 orders. The photoresponsive behavior of nematic E7 doped with TR5 was investigated. This TR5–E7 system showed a large factor for dye-induced optical torque enhancement. To enhancement of the photoresponse of the oligothiophene derivatives, ester moieties were introduced into their molecules and their ability to reorient host LMWLCs with regard to the absorption, the dipole–dipole interaction with the LC molecules, and the intramolecular delocalization of π-electrons was studied [Yaegashi et al., 2005]. LC thiophene derivatives having ester moieties directly connected to the terminal thiophene (TD1) lowered the reorientation threshold intensities by about three times compared to that of the thiophene dye without ester moiety at the same dye concentration. Even at the same absorbance at 488 nm, the TD1-doped LC cells showed lower threshold intensity than cells doped with indirectly connected to the terminal thiophene (TD2).

As one of the applications of the Jánossy effect of LMWLCs, planar microlenses and microlens arrays were fabricated by combination of photophysically induced reorientation of dye-doped LC/monomer mixtures with their simultaneous photopolymerization [Yaegashi et al., 2007]. As shown in Fig. 4.31, observation of the microlens revealed that the molecular alignment of LCs induced by the actinic beam remained unchangeable after turning the beam off. Each microlens in the arrays exhibited polarization selectivity with respect to the polarization direction of the incident light, and was arranged in any desired direction. The microlens array developed by this method is one of promising candidates for broad applications, such as, key components in optical parallel-processing systems and large-scale free-space networks.

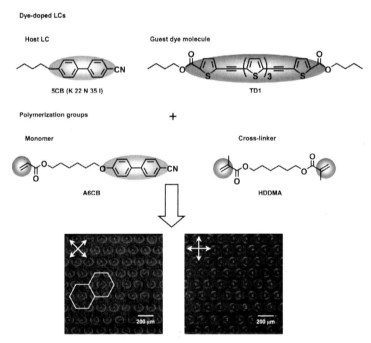

Figure 4.31 Microlens array fabricated by combination of photoinduced reorientation of dye-doped LC/monomer mixtures and simultaneous photopolymerization.

Unlike photochemical and photophysical effects for order–order transition processes, photothermal effect induces an order–disorder transition, which brings about random arrangement of LMWLCs, as shown in Fig. 4.27. Such an effect is the simplest effect connected with light absorption. In general, LMWLCs show large temperature gradients of refractive indices, dn_e/dt and dn_o/dt. The changes in LC alignment are observed due to light absorption and heating of the LCs. For instance, n_e of 5CB decreases from 1.682 to 1.635, and n_o of 5CB increases from 1.517 to 1.532 with an increase in temperature from 23 to 35°C, respectively. This temperature dependence of n_e and n_o may be interpreted in terms of the change in an order parameter of the LC and the change in density.

4.6 Photodriven Motion of LMWLCs

For photochromic molecules such as azobenzenes, their photoreaction often induces a large change in molecular shape. However, such photoresponse is limited in a molecular scale, which cannot be observed directly. In photoresponsive LMWLC systems containing photoresponsive LCs, the photoinduced deformation can be enlarged or enhanced by MCM because of the inherent self-organization of LCs. Thus, this kind of photoresponse can be conveniently characterized.

Recently, one azobenzene compound showing nematic LC phase at room temperature was synthesized [Okano et al., 2009]. The photoinduced expansion and condensation behavior in this azobenzene LCs dispersed on water surface, as shown in Fig. 4.32. The photoinduced expansion was observed with pure azobenzene LMWLCs, which was attributed to both photoisomerization and photoinduced LC-to-isotropic phase transition. On the other hand, photoinduced condensation behavior was obtained in the LMWLC/DMF mixture. This contrastive effect is related with the interfacial tension.

Molecular motors can photodrive visible motions of micro-objects on the surface of CLC film by photoinduced reversible change in LC texture [Vicario et al., 2006]. As shown in Fig. 4.33, the rotated glass rod exceeds the size of the motor molecule by a factor of 10,000. The changes in shape of the motor during the rotary steps cause a remarkable rotational reorganization of the cholesteric LC film and its surface relief verified with the atomic force microscopy (AFM) images in taping mode.

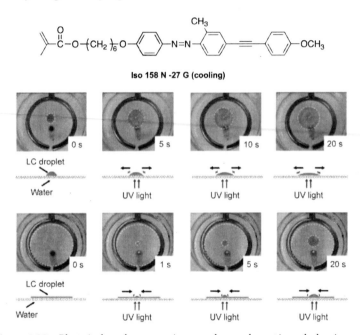

Figure 4.32 Photoinduced expansion and condensation behavior of azobenzene LMWLC dispersed on water surface. Reprinted with permission from Okano et al., Copyright 2009, WILEY-VCH Verlag GmbH & Co. KGaA, Weinheim.

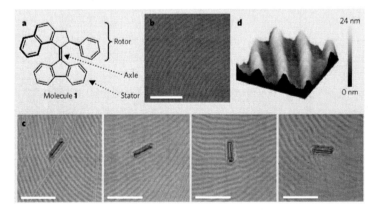

Figure 4.33 Photoinduced motion of glass rod embedded in CLC film doped with one molecular motor.

Kausar et al. [2009] reported a visible photodriven motion of polystyrene (PS) microparticles dispersed on the surface of LMWLC films. As shown in Fig. 4.34, the PS microparticle moved toward the irradiation position upon UV irradiation. Visible light irradiation drove the PS microparticle move in an opposite direction. The speed of the motion can be controllable by changing the light intensity or the concentration of the azobenzene compound in LMWLCs. Using glassy micro-rod; both rotational and translational motions were obtained depending on the phases of the LMWLCs [Kausar et al., 2011].

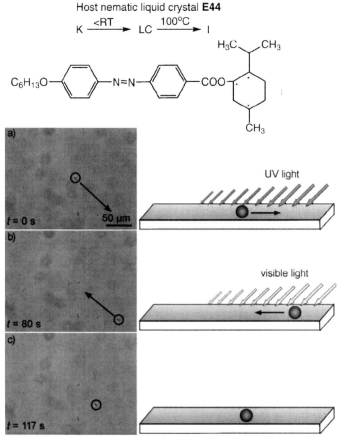

Figure 4.34 Photodriven motion of PS microparticles dispersed on the surface of films of azobenzene-containing LMWLCs. Reprinted with permission from Kausar et al., Copyright 2009, WILEY-VCH Verlag GmbH & Co. KGaA, Weinheim.

References

Allinson, H. and Gleeson, H. F. (1993). Physical properties of mixtures of low molar mass nematic liquid crystals with photochromic fulgide guest dyes, *Liq. Cryst.* **14**, pp. 1469–1478.

Allinson, H. and Gleeson, H. F. (1995). Variations in the elastic constants of a fulgide-doped liquid crystal system, *J. Mater. Chem.* **5**, pp. 2139–2144.

Alam, M. Z., Yoshioka, T., Ogata,T., Nonaka,T., and Kurihara, S. (2007). Influence of helical twisting power on the photo switching behavior of chiral azobenzene compounds: Applications to high-performance switching devices, *Chem. Eur. J.* **13**, pp. 2641–2647.

Denekamp, C. and Feringa, B. L. (1998). Optically active diarylethenes for multimode photoswitching between liquid-crystalline phases, *Adv. Mater.* **10**, pp. 1080–1082.

Durbin, S. D., Arakelian, S. M., and Shen, Y. R. (1981). Optical field-induced birefringence and Freedericksz transition in a nematic liquid crystal, *Phys. Rev. Lett.* **47**, pp. 1411–1414.

Eelkema, R. and Feringa, B. L. (2006). Reversible full-range color control of a cholesteric liquid-crystalline film by using a molecular motor, *Chem. Asian J.* **1**, pp. 367–369.

Gibbons, W. M., Shannon, P. J., Sun, S. T., and Swetlin, B. J. (1991). Surface-mediated alignment of nematic liquid crystals with polarized laser light, *Nature* **351**, pp. 49–50.

Hu, W., Zhao, H., Song, L., Yang, Z., Cao, H., Cheng, Z., Liu, Q., and Yang, H. (2010). Electrically controllable selective reflection of chiral nematic liquid crystal/chiral ionic liquid composites, *Adv. Mater.* **22**, pp. 468–472.

Ikeda, T., Sasaki, T., and Ichimura, K. (1993). Photochemical switching of polarization in ferroelectric liquid-crystal films, *Nature* **361**, pp. 428–430.

Jánossy, I., Lloyd, A. D., and Wherrett, B. S. (1990). Anomalous optical Freederickes transition in an absorbing liquid crystal, *Mol. Cryst. Liq. Cryst.* **179**, pp. 1–12.

Kajiyama, T., Kikuchi, H., and Nakamura, K. (1993). Photoresponsive electro-optical effect of (liquid crystalline polymer)/(low molecular weight liquid crystal) composite system, *Proc. IS&T/SPIE's Symposium on Electric Imaging* **1911**, pp. 111–121.

Kausar, A., Nagano, H., Ogata, T., Nonaka, T., and Kurihara, S. (2009). Photocontrolled translational motion of a microscale solid object on azobenzene-doped liquid-crystalline films, *Angew. Chem. Int. Ed.* **48**, pp. 2144–2147.

Kausar, A., Nagano, H., Kuwahara, Y., Ogata, T., and Kurihara, S. (2011). Photocontrolled manipulation of a microscale object: A rotational or translational mechanism, *Chem. Eur. J.* **17**, pp. 508–515.

Kreuzer, M., Hanisch, F., Eidenschink, R., Paparo, D., and Marrucci, L. (2002). Large deuterium isotope effect in the optical nonlinearity of dye-doped liquid crystals, *Phys. Rev. Lett.* **88**, pp. 013902 (1–4).

Kurihara, S., Ikeda, T., Tazuke, S., and Seto, J. (1991). Isothermal phase transition of liquid crystals induced by photoisomerization of doped spiropyrans, *J. Chem. Soc. Faraday Trans.* **87**, pp. 3251–3254.

Kumar, G. S. and Neckers, D. C. (1989). Photochemistry of azobenzene-containing polymers, *Chem. Rev.* **89**, pp. 1915–1925.

Lee, H. K., Doi, K., Harada, H., Tsutsumi, O., Kanazawa, A., Shiono, T., and Ikeda, T. (2000). Photochemical modulation of color and transmittance in chiral nematic liquid crystals containing an azobenzene as a photosensitive chromophore, *J. Phys. Chem. B* **104**, pp. 7023–7028.

Lee, H. K., Kanazawa, A., Shiono, T., Ikeda, T., Fujisawa, T., Aizawa, M., and Lee, B. (1998). All optically controllable polymer/liquid-crystal composite films containing azobenzene liquid crystal, *Chem. Mater.* **10**, pp. 1402–1407.

Li, Q., Li, Y., Ma, J., Yang, D., White, T. J., and Bunning, T. J. (2011). Directing dynamic control of red, green, and blue reflection enabled by a light-driven self-organized helical superstructure, *Adv. Mater.* **23**, pp. 5069–5073.

Mamiya, J., Kanie, K., Hiyama, T., Ikeda, T., and Kato, T. (2002). A rodlike organogelator: Fibrous aggregation of azobenzene derivatives with a syn-chiral carbonate moiety, *Chem. Commun.* pp. 1870–1871.

Moriyama, T., Kajita, J., Takanashi, Y., Ishikawa, K., Takezoe, H., and Fukuda, A. (1993). Optically addressed spatial light modulator using an antiferroelectric liquid crystal doped with azobenzene, *Jpn. J. Appl. Phys.* **32**, pp. L589–L592.

Moriyama, M., Mizoshita, N., Yokota, T., Kishimoto, K., and Kato, T. (2003). Photoresponsive anisotropic soft-solids: Liquid-crystalline physical gels based on a chiral photochromic gelator, *Adv. Mater.* **15**, pp. 1335–1338.

Natansohn, A. and Rochon, P. (2002). Photoinduced motions in azo-containing polymers, *Chem. Rev.* **102**, pp. 4139–4175.

Okano, K., Shinohara, M., and Yamashita, T. (2009). Light-induced deformation of photoresponsive liquid crystals on a water surface, *Chem. Eur. J.* **15**, pp. 3657–3660.

Sackmann, E. (1971). Photochemically induced reversible color changes in cholesteric liquid crystals, *J. Am. Chem. Soc.* **93**, pp. 7088–7090.

Sasaki, T., Ikeda, T., and Ichimura, K. (1994). Photochemical control of properties of ferroelectric liquid crystals: Photochemical flip of polarization, *J. Am. Chem. Soc.* **116**, pp. 625–628.

Shirota, K. and Yamaguchi, I. (1997). Optical switching of antiferroelectric liquid crystal with azo-dye using photochemically induced SmC A*-SmC* phase, *Jpn. J. Appl. Phys.* **36**, pp. L1035–L1037.

Sung, J., Hirano, S., Tsutsumi, O., Kanazawa, A., Shiono, T., and Ikeda, T. (2002). Dynamics of photochemical phase transition of guest/host liquid crystals with an azobenzene derivative as a photosensitive chromophore, *Chem. Mater.* **14**, pp. 385–391.

Takizawa, K., Kikuchi, H., Fujikake, H., Namikawa, Y., and Tada, K. (1994). Reflection mode polymer-dispersed liquid crystal light valve, *Jpn. J. Appl. Phys.* **33**, pp. 1346–1351.

Tamaoki, N. (2001). Cholesteric liquid crystals for color information technology, *Adv. Mater.* **13**, pp. 1135–1147.

Tazuke, S., Kurihara, S., and Ikeda, T. (1987). Amplified image recording in liquid crystal media by means of photochemically triggered phase transition, *Chem. Lett.* pp. 911–914.

Tong, X. and Zhao, Y. (2003). Self-assembled cholesteric liquid crystal gels: Preparation and scattering-based electrooptical switching, *J. Mater. Chem.* **13**, pp. 1491–1495.

Vicario, J., Katsonis, N., Ramon, B. S., Bastiaansen, C. W. M., Broer, D. J., and Feringa, B. L. (2006). Nanomotor rotates microscale objects, *Nature* **440**, pp. 163–163.

White, T. J., Freer, A. S., Tabiryan, N. V., and Bunning, T. J. (2010). Photoinduced broadening of cholesteric liquid crystal reflectors, *J. Appl. Phys.* **107**, pp. 073110 (1–6).

White, T. J., Bricker, R. L., Natarajan, L. V., Tabiryan, N. V., Green, L., Li, Q., and Bunning, T. J. (2009). Phototunable azobenzene cholesteric liquid crystals with 2000 nm range, *Adv. Funct. Mater.* **19**, pp. 3484–3488.

Yamaguchi, T., Inagawa, T., Nakazumi, H., Irie, S., and Irie, M. (2000). Photoswithing of helical twisting power of a chiral diarylethene dopant: Pitch change in a chiral nematic liquid crystal, *Chem. Mater.* **12**, pp. 869–871.

Yaegashi, M., Shishido, A., Shiono, T., and Ikeda, T. (2005). Effect of ester moieties in dye structures on photoinduced reorientation of dye-doped liquid crystals, *Chem. Mater.* **17**, pp. 4304–4309.

Yaegashi, M., Kinoshita, M., Shishido, A., and Ikeda, T. (2007). Direct fabrication of microlens arrays with polarization selectivity, *Adv. Mater.* **19**, pp. 801–804.

Yang, D. K., Huange, X. Y., and Zhu, Y. M. (1997). Bistabel cholesteric reflective displays: Materials and drive schemes, *Annu. Rev. Mater. Sci.* **27**, pp. 117–146.

Zhang, H., Shiino, S., Shishido, A., Kanazawa, A., Tsutsumi, O., Shiono, T., and Ikeda, T. (2000). A thiophene liquid crystal as a novel p-conjugated dye for photo-manipulation of molecular alignment, *Adv. Mater.* **12**, pp. 1336–1339.

Zhao, Y. and Tong, X. (2003). Light-induced reorganization in self-assembled liquid crystal gels: Electrically switchable diffraction gratings, *Adv. Mater.* **15**, pp. 1431–1435.

Chapter 5

Liquid Crystal Polymers

With birefringent mesogens in side chain or main chain, LC polymers (LCPs) integrate LC properties with high-performance polymer materials possessing a film-forming nature, high processability, easy-fabrication characteristics, high corrosion resistance, and low manufacturing costs. Introduction of photoresponsive mesogens into LCPs provides the designed materials with photocontrolled features, as shown in Fig. 5.1. Among them, photochemical phase transition and photocontrolled alignment are the most important ones.

Therefore, the photoresponsive LCPs have been widely explored as promising photonic materials due to their controllable properties in response to light. For photoresponsive features, side-chain LCPs are usually superior to main-chain ones because the latter usually exhibits lack of molecular mobility by confining mesogens in macromolecular main chain. Scheme 5.1 shows synthesis of one typical nematic LCP with photoisomerizable azobenzenes in side chain as nematic mesogens.

5.1 Photochemical Phase Transition

5.1.1 Copolymers and Polymer Composites

In 1987, light induced a change in LC copolymers containing azobenzene moieties and photoinert mesogenic groups for

---

Dancing with Light: Advances in Photofunctional Liquid-Crystalline Materials
Haifeng Yu
Copyright © 2015 Pan Stanford Publishing Pte. Ltd.
ISBN 978-981-4411-11-0 (Hardcover), 978-981-4411-12-7 (eBook)
www.panstanford.com

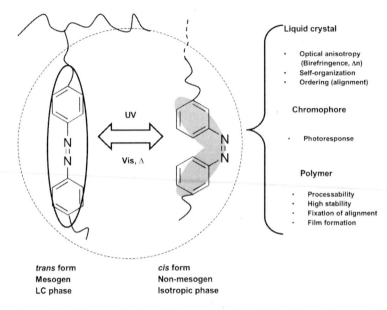

Figure 5.1 Characteristics of photoresponsive LCPs with azobenzene in the side chain.

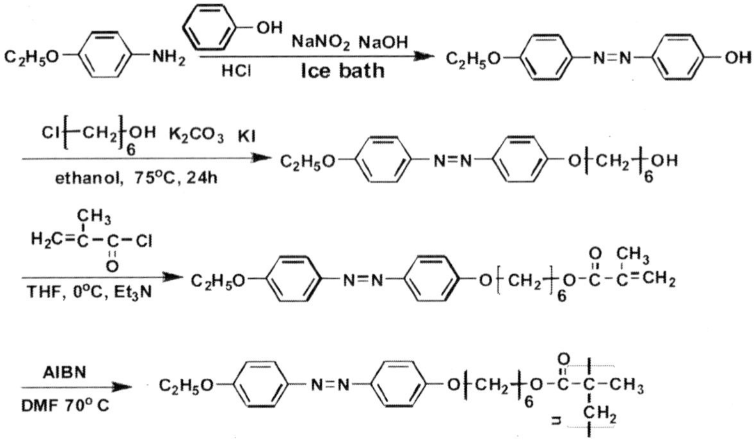

Scheme 5.1 Synthetic route of one typical nematic LCP.

photonic applications was first reported [Eich et al., 1987; Eich & Wendorff, 1987]. Ikeda et al. [1988] reported the photochemical phase transition in low-molecular-weight (LMW) azobenzene/LCP guest–host composites. As shown in Fig. 5.2, they demonstrated that the photochromic composites underwent a nematic LC-to-isotropic phase transition upon photoirradiation to cause *trans–cis* isomerization of azobenzenes, and the isotropic mixture reverted to the initial nematic LC phase with *cis–trans* back-isomerization being irradiated with visible light.

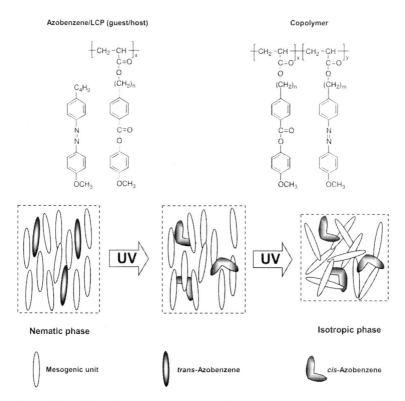

Figure 5.2 Photochemical phase transition in azobenzene/LCP or LC copolymer systems.

Although the first example of the photochemical phase transition was demonstrated in the LMW azobenzene/LCP guest–host systems, it became obvious that LC copolymers are more advantageous than composites of LMW azobenzene/LCP guest–host mixtures, because of the low solubility of the guest dye molecules in polymer matrix. Furthermore, in some cases phase separation occurs when the concentration of the guest molecules is high, which decreases the optical performances of materials. Thus, the studies on dye-doped LCP composites were soon extended to polymers containing photochromic molecules, showing advanced performance of the photochemical phase transition [Ikeda et al., 1990]. In addition to azobenzenes, fulgide, menthone, spiropyran, and cyanophenyl benzoate moieties have been explored in LCPs and copolymers.

Generally, the photonic applications of LCPs originate from their photomanipulation of optical features by actinic light, as shown in Fig. 5.3. There are many factors influencing the properties of LCPs, and the responsive speed in the optical processing is one of the most important parameters among them. In one mixture of azobenzene and LMWLCs, the nematic LC-isotropic phase transition occurred in 100 ms as indicated by the loss of birefringence of the sample films. The nematic LC-isotropic phase transitions in azobenzene/LCP composites were also found to take place in 50–200 ms in azobenzene-doped LCPs and azobenzene-containing LC copolymers.

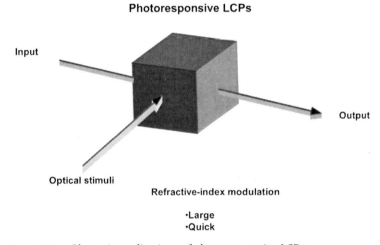

Figure 5.3 Photonic applications of photoresponsive LCPs.

5.1.2 Homopolymers

In the process of photochemical phase transition of LC copolymers, a relatively longer time is necessary for orientational relaxation of mesogens because of their higher viscosity comparing with LMWLCs. To overcome this difficulty, a new system has been developed, in which each mesogen in LCPs is provided with a photoresponsive moiety [Ikeda & Tsutsumi, 1995]. For instance, in azobenzene-containing LCPs, the azobenzene moiety could play both roles as a mesogen and as a photoresponsive group when it is connected with polymer main with a soft spacer (Fig. 5.4).

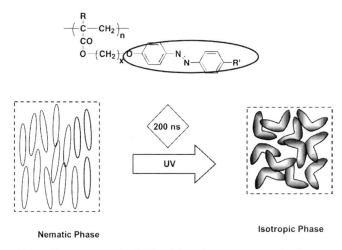

Figure 5.4 Photoresponsive LCPs with azobenzenes act as both mesogenic moieties and light-responsive groups.

Figure 5.5 gives one typical example of photoresponsive LCPs with azobenzene moieties in the side chain (PM6ABOC2). It shows a very stable nematic LC phase between the glass transition temperature (T_g) of the LCP and the clearing point (about 150°C). These azobenzenes show an LC phase only when they are in the *trans* forms, and they never show any LC phases in their *cis* forms.

Since the photoisomerizable azobenzene group is the only mesogen in the LCP, it was observed in Fig. 5.6 that photochemical phase transition was essentially induced on the same timescale as photochemical reactions of the photoresponsive moiety in each mesogen, when the photochemical reactions of a large number of mesogens are induced simultaneously by means of a short laser pulse. It was found that the nematic LC-isotropic phase transition was obtained in thin films of PA6ABOC2 in 200 ns upon irradiation with a laser pulse (20 ps) [Kanazawa et al., 1997]. The change in refractive index was observed at about 0.1 within such a short time. This quick photoresponse of LCPs is an encouraging result from the viewpoint of applications in photonic devices.

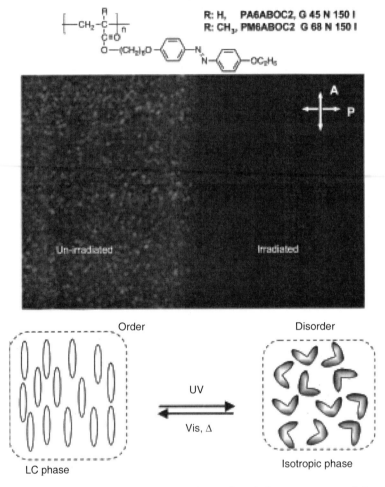

Figure 5.5 One typical example of photoinduced phase transition of thin films of one photoresponsive LCP (PM6ABOC2).

Furthermore, the photochemical phase transition in LCPs with an azobenzene as the only one mesogen can be induced in an extremely wide temperature range from room temperature to over 150°C. This advantageous feature as photonic materials is characteristic to photoresponsive LCPs, in which each mesogen includes a photoactive moiety. To develop high-performance photonic materials, the photochemical phase-transition behavior of azobenzene-containing LCPs has been extensively explored with a special reference to the structure of LCPs. In such photonic applications as all-optical switching and dynamic holography, it is necessary to rapidly induce not only LC-to-isotropic phase transition, but also isotropic-to-LC phase transition (recovery of the initial phase). When photoirradiation is stopped after the LC-isotropic phase transition is induced, the initial LC phase is restored after some time because of thermal *cis–trans* back-isomerization due to the thermal instability of *cis*-azobenzenes. Generally, the relaxation time is strongly dependent on the experimental temperature and the kinds of azobenzenes described in Chapter 3. Moreover, the structure of polymer main chain also has a great influence on the performance of photoresponsive LCPs.

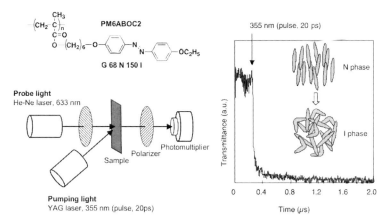

Figure 5.6 Time-resolved measurement of quick response by means of photoinduced phase transition in thin films of one photoresponsive LCP with a photoisomerizable azobenzene group as the only mesogen.

To enhance the thermal recovery, the mechanism of the isotropic-nematic LC phase transition was investigated in detail. The recovery is composed of two processes: thermal *cis–trans* back-isomerization of the azobenzene moieties and reorientation of the mesogenic *trans*-azobenzenes, and it was observed that the *cis–trans* back-isomerization is a rate-determining process [Tsutsumi et al., 1988]. As shown in Fig. 5.5, it was also found that polymethacrylate with side-chain azobenzenes (PM6ABOC2) possess a higher T_g than the corresponding polyacrylate with the same side-chain structures (PA6ABOC2), and azobenzene groups in both of the LCPs exhibit a similar photoresponse: they showed similar photochemical nematic LC-to-isotropic phase transition behavior. However, the recovery of the initial nematic LC phase was much slower in the PM6ABOC2 than in PA6ABOC2 even though the thermal *cis–trans* back-isomerization takes place at a similar rate. Obviously, this is due to the stiffness of the main chain of PM6ABOC2, which significantly affects the reorientation process of the *trans*-azobenzene mesogens after *cis–trans* back-isomerization. Furthermore, on the basis of the results of the kinetic studies on the isotropic-to-nematic LC phase transition, azobenzene-containing LCPs with both donor and acceptor moieties in one molecule, characterized by very fast *cis–trans* thermal back-isomerization, have been designed. With such donor–acceptor azobenzenes, a very fast recovery of the nematic phase (800 ms) was achieved. This response is faster than that of conventional azobenzene LCs by one order of magnitude. The effect of the spacer between the polymer main chain and the side chain of the azobenzene chromophore on the photochemical phase transition behavior was explored. It was found that the spacer strongly affects the photoresponsive behavior.

Similarly, a novel reflection mode was developed, which also shows a fast photochemical phase transition in azobenzene-containing LC materials. On laser pulse irradiation, it is possible to switch the incident probe light reflected from the interface between LCs and a substrate, as a result of modulation of reflectivity arising from a photoinduced change in the refractive index of LC materials. In one LMWLC (BMAB in Fig. 5.7), the reflection-mode system gave a similar response to that observed in the usual transmission-mode systems; however, it gave a decay time of 1 ms, which is significantly shorter than that obtained in the transmission-mode systems (Fig. 5.7). The molar extinction coefficient of the azobenzene moiety is very large ($>10^4$) at 355 nm. Thus, actinic light is absorbed entirely at the surface of the sample. As a result, *trans–cis* photoisomerization is also induced near the surface, so that the nematic LC-to-isotropic phase transition occurs only in the surface region, leaving the bulk area intact as a nematic phase. In the reflection-mode systems, the probe light can only penetrate the surface area, so if molecules in the *cis* form produced at the surface by pulse irradiation diffuse into the bulk phase, and simultaneously they are replaced by molecules in the *trans* form from the bulk phase, recovery of the initial nematic phase can be achieved without involvement of a slow *cis–trans* back-isomerization process. Since diffusion and reorientation processes are much faster than *cis–trans* back-isomerization, the reflection-mode optical switching has thus become much faster (Fig. 5.7).

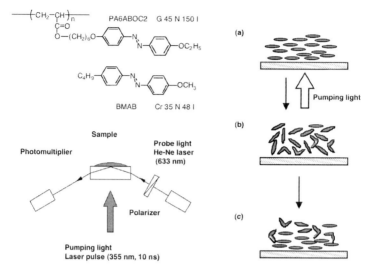

Figure 5.7 Reflection-mode optical switching of azobenzene LC materials and its possible mechanism (a) nematic LC phase (b) isotropic phase induced by light (c) nematic LC phase thermally recovered. Reprinted with permission from Shishido et al., Copyright 1997, American Chemical Society.

Such reflection-mode systems have another superior characteristic for optical switching. For the practical use of optical switching devices containing organic dyes as a key component, there is a very important prerequisite, the stability. Optical switching in the ordinary transmission-mode generally exhibits low fatigue resistance. However, the reflection-mode optical switching was found to be repeatable over 15,000 cycles, which is almost 10 times more fatigue-resistant than the conventional transmission-mode switching. This suggests that optimization of this optical system for photoresponsive LCs may be an effective approach to realize more stable optical switching devices [Shishido et al., 1997].

In 1995, the application of azobenzene-containing LCPs as optical image storage materials was first reported. In a nematic glassy state, LCPs undergo a nematic LC-to-isotropic phase transition upon photoirradiation while they never exhibit an isotropic-to-nematic LC phase transition below their T_gs. Figure 5.8 shows photographs of a photomask and a binary test pattern recorded in LCP films at room temperature (below T_g) by pulse irradiation of UV light at 355 nm.

It is obvious that the irradiated area becomes isotropic as evidenced by the loss of birefringence upon observation of a polarizing optical microscope (POM). The stored optical image has been kept stable over several years at room light. In the LCP films, it was observed that the thermal *cis–trans* back-isomerization of azobenzene mesogens took place in 24 h at room temperature. Although the *trans* form was recovered completely, the isotropic glassy domains induced at the irradiated area still remained unchangeable at room temperature even after several years. These results suggest that the orientation of the mesogenic *trans*-azobenzenes becomes disordered below T_g through thermal *cis–trans* back-isomerization process. However, even after the *trans* form is thermally recovered, the alignment of mesogens is difficult in the absence of segmental motions of the main chain of the LCP below T_g. The recorded images can be deleted by heating the sample films higher than its T_g, which can be used for rewritable storage of new images. Therefore, the photoresponsive LCP can be used as optical switching as well as optical image storage materials as shown Fig. 5.8.

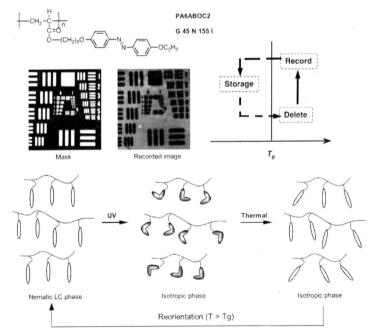

Figure 5.8 Optical image storage and data processing in PA6ABOC2. The recorded image was taken upon observation with POM.

5.2 Photoinduced Cooperative Motion

As described in Chapter 3, photoresponsive LCs can be aligned upon irradiation of linearly polarized actinic light with suitable wavelength. After this photoalignment, a large anisotropy can be induced accordingly. There have been many studies on the photoinduced anisotropy in alignment of LC copolymer systems with both mesogens and azobenzene moieties in the side chains. Several studies clearly showed that non-photoactive mesogens could undergo reorientation concomitantly with azobenzene moieties above T_g owing to their molecular cooperative motion (MCM), as shown in Figs. 5.9 and 5.10. In addition, different arguments were presented concerning reorientation behavior below T_g. Recent researches on the photoalignment behavior of side-chain LCPs have demonstrated that dipole–dipole interactions affecting molecular cooperative motion is crucial in the reorientation process.

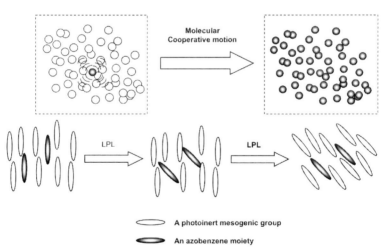

Figure 5.9 Photoinduced MCM in photoresponsive LCPs.

In the process of reorientation of azobenzene-containing LCPs, several important factors have been intensively studied, including the azobenzene contents, enthalpy changes during phase transitions, morphology of samples before irradiation, intensity of actinic light [Wu et al., 1998a], and spacer length of side-chains in LCPs [Wu et al., 1998b]. Besides, structural effects of azobenzene moieties on the photoalignment behavior have been systematically explored in a series of side-chain LCPs with different azobenzene moieties [Wu et al., 1998a; 1998c]. With an increase in the strength of donor and acceptor substituents at 4- and 4′-positions of the azobenzene moieties, the possibility of alignment change in LCPs decreased due to slightly increased enthalpy stability in an LC phase and a significantly reduced concentration of *cis*-azobenzene (i.e., an increased *cis–trans* thermal isomerization rate). However, high alignment efficiency was observed in a polyacrylate with strong donor–acceptor pairs in the azobenzene moiety and exhibiting a low stability of LC phases, since both the rate of *cis–trans* isomerization and the mobility of mesogens are favorable for the alignment change. Interestingly, an azobenzene LCP possessing a smectic LC phase showed in-plane alignment when it was grafted onto surface of silicon wafers, which was found to be more sensitive to undergo photoalignment with linearly polarized light (LPL) [Uekusa et al., 2009]. These results indicate that the photoalignment behavior can be effectively optimized by an appropriate choice of azobenzene units and polymer backbones.

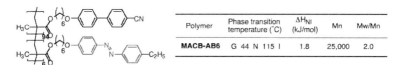

Polymer	Phase transition temperature (°C)	ΔH_{NI} (kJ/mol)	Mn	Mw/Mn
MACB-AB6	G 44 N 115 I	1.8	25,000	2.0

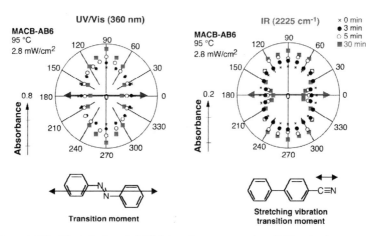

Figure 5.10 Photoinduced MCM in photoresponsive LCPs. Azobenzene, the photoresponsive mesogen and cyanobiphenyl group, the photoinert one.

5.3 Photoinduced Large Change in Birefringence

As shown in Fig. 5.3, photoresponsive LCPs are suitable for photonic applications beneficial from their quick optical process and large modulation of refractive index. It is well known that optical birefringence can be produced by change in transition moments of chromophores by photoalignment. Early work in this field was carried out on LPL-induced change in alignment of azobenzenes embedded in amorphous polymers. The first report about the generation of birefringence was employed with azobenzene-doped films of polyvinyl alcohol (PVA). Although many researches focused on the chromophore-doped polymers, the stability of the induced anisotropy was low. In amorphous polymers covalently attached with a chromophore moiety, optical anisotropy was also observed upon LPL irradiation. These polymers showed stable photoinduced birefringence at an ambient temperature below their T_gs. The induced birefringence could be erased thermally and photochemically, which can be regenerated upon LPL irradiation. But the formed birefringence is small due to the amorphous state, which cannot meet the requirement of photonic applications.

A large change in birefringence can be induced in azobenzene-containing LCPs by manipulating the LC alignment, which is more efficient than that of amorphous polymers. Moreover, the alignment of azobenzenes can be photomanipulated in a 3D way, whose ordering can be then transferred to other photoinactive mesogens by molecular cooperative motion. In other words, a large and stable birefringence can be photoinduced than that obtained from materials with an azobenzene as one single mesogen. It is well known that tolane is one of the most common core structures for design of highly birefringent LC molecules because of its longer molecular conjugation length [Sekine et al., 2001]. Using copolymerization method, LC copolymers containing both tolane and azobenzene moieties in the side chain have been developed, and a large change in birefringence was obtained in a homogeneously aligned state [Yoneyama et al., 2002]. However, the photosensitivity was not high enough due to a low content of azobenzene moieties in the copolymers. To increase the photoresponsive speed and induce a larger birefringence for photonic applications, they proposed a concept of molecular architecture of azobenzene-containing LC materials. As shown in Fig. 5.11, a tolane group is directly attached onto the 4- or 4'- position of an azobenzene molecule to prepare an azotolane mesogen, which shows far longer molecular conjugation length than one single tolane group or an azobenzene moiety. Moreover, both the photoresponse and the MCM are greatly enhanced in the novel system, and a large photoinduced change in birefringence is expected.

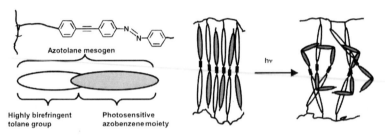

Figure 5.11 Molecular architectures of LCPs containing an azotolane mesogen for a large photoinduced change in birefringence.

Figure 5.12 lists the molecular structures and LC phase behaviors of a series of azotolane LCPs synthesized based on this principle. Here, three, four, and five benzene rings are included in the mesogens, respectively [Okano et al., 2006a, 2006b, 2006c]. All the prepared azotolane LCPs showed a wide nematic LC range and exhibited a phase transition temperature higher than 200°C. In general, a nematic LC phase is quite advantageous for photonic applications because its sensitivity to external stimuli is higher than that of other LC phases. Upon irradiation with actinic light, a huge birefringence larger than 0.35 at 633 nm was obtained by the photochemical phase transitions of the azotolane materials with three benzene rings (P3AT, P3TA, and P3AT-NO$_2$). The position of the tolane moiety in the azotolane group and that of the donor–acceptor substituent showed great influence on the photochemical phase transitions and the photoinduced change in birefringence. Although the cis–trans back-isomerization of the LCP with donor–acceptor azotolane moieties (P3AT-NO$_2$) in solution was faster than that of LCPs having no donor–acceptor groups, the P3AT-NO$_2$ films still showed a little change in birefringence upon UV irradiation. Furthermore, the large birefringence can be obtained even at the telecommunication wavelength (1550 nm) by the photoinduced phase transition from a nematic LC to an isotropic state [Okano et al., 2006c].

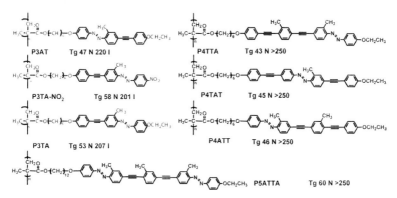

Figure 5.12 Typical examples of LCPs containing an azotolane mesogen for a large photoinduced change in birefringence.

A larger birefringence (>0.65 at 633 nm) was obtained for the azotolane LCPs with a longer molecular conjugation length (four benzene rings, P4TTA, P4TAT, P4ATT) by photoinduced change in alignment of azotolane mesogens. The largest birefringence of about 0.76 was obtained for the LCPs possessing the longest molecular length of an azotolane mesogen with five benzene ring (P5ATTA) [Okano et al., 2006b]. P5ATTA with two azobenzene units in one azotolane mesogen also showed the most efficient change in alignment, indicating that the high birefringence in the homogeneously aligned state could be efficiently converted to a change in birefringence.

Following the molecular design of azotolane materials, a large birefringence of about 0.3 was also induced by the method of anisotropic photocrosslinking and thermal treatment of LCPs with both a tolane and a cinnamate group in the same mesogen [Kawatsuki et al., 2008]. Moreover, the photoinduced birefringence was precisely controlled by post functionalization [Yu et al., 2009]. As shown in Fig. 5.13, functionalized LCPs with different degree of functionality were synthesized via post Sonogashira cross-coupling reaction of a polymer precursor.

The post-functionalization was easily carried out in a mild condition and showed a high yield, as shown in Fig. 5.14. Although a highly birefringent azotolane group was introduced into the polymer precursor, the photoresponse of the functionalized LC materials was not obviously decreased. By adjusting the content of azotolane groups, precise control of the photoinduced birefringence was successfully obtained. The present method of precise control of photoinduced birefringence enables one to finely photocontrol optical performances of materials, indicating its potential applications as advanced processing for photonic materials.

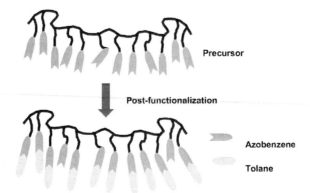

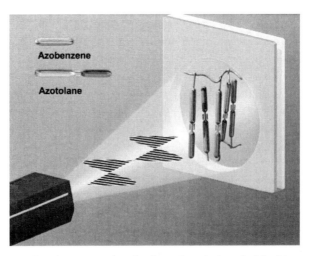

Figure 5.13 Precise control of the photoinduced birefringence in photoresponsive LCP by post-functionalization. Reprinted with permission from Yu et al., Copyright 2009, WILEY-VCH Verlag GmbH & Co. KGaA, Weinheim.

Recently, azotolane materials were reported to show an LC phase at room temperature when oligo (oxyethylene) groups are connected into azotolane mesogens [Okano et al., 2008]. The molecular architecture of azotolane LCPs with a giant photoinduced change in birefringence is useful for extensive optical applications in high performance photonic devices (such as high-density optical recording, as well as holographic and multi-bit recording) and photoswitching materials.

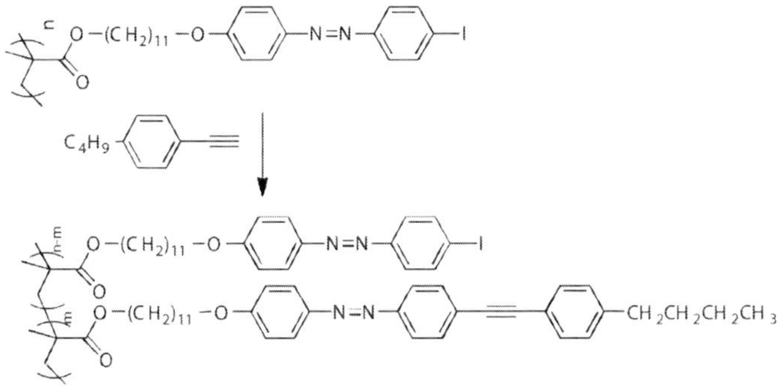

Figure 5.14 Molecular design of the post-functionalization in Fig. 5.13. Reprinted with permission from Yu et al., Copyright 2009, WILEY-VCH Verlag GmbH & Co. KGaA, Weinheim.

5.4 Polarized Electroluminescence (EL)

Recently, electroluminescence (EL) materials based on conjugated polymers exhibiting polarized emission behaviors have attracted much attention because of their potential applications as backlights for conventional LCDs. The introduction of molecular alignment into emissive layers is one of the most promising strategies for an elaborate control of emission properties in thin films. The approach to such a device is based on the use of organic materials that are optically anisotropic and can be aligned along one specific direction, as shown in Fig. 5.15.

The strong anisotropy of LC materials and their self-organization ability enable one to easily fabricate well-oriented thin films [Grell & Bradley, 1999], and integration of LC properties into EL materials may bright about unique properties. For example, the self-organized structures are controlled by the choice of phases of LC materials, so that structurally controlled thin films could be prepared by quenching from appropriately ordered phases into solid states. Due to these properties, EL materials with LC features are expected to be applied not only in display devices but also in various photonic applications.

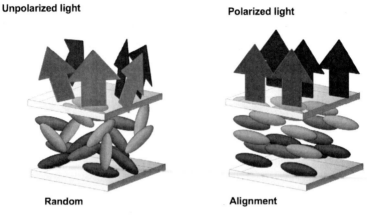

Figure 5.15 Schematic illustration of polarized EL based on alignment of LCPs.

In preparation of EL devices, multi-layer devices were found to be more efficient than single-layer ones, but the latter can be fabricated much more simply. As a result, mixtures composed of hole transporters, electron transporters, and emitters, were often used to prepare efficient single-layer devices. However, the microphase separation occurring in different components reduces the device lifetimes. From the viewpoint of molecular design, many efforts have been attempted to develop novel materials with improved electron-transporting ability for fabrication of single-layer devices. In 1998, an amorphous polymer with bipolar charge-transporting ability was explored for single-layer EL devices, which revealed good operating stabilities and external quantum efficiency up to 0.15% with an aluminum electrode [Peng et al., 1998]. But they did not report the anisotropic EL properties. Then, a LMW compound composed of an oxadiazole (OXD) moiety as an electron-transporting unit and an amine moiety as a hole-transporting unit was demonstrated to show polarized EL properties in anisotropic LC host [Mochizuki et al., 2000a]. They introduced a biphenyl group as a mesogen into the obtained bipolar materials to allow the appearance of LC phases for anisotropic control [Mochizuki et al., 2000b].

To improve the processability, an LCP (PM6-OXD-MA) shown in Fig. 5.16 was prepared, in which the bipolar carrier-transporting groups was designed as side chain [Mochizuki et al., 2003]. In this case, an alignment layer on the surface of indium tin oxide (ITO)-

coated glass substrates was used to align LC molecules and the polarized EL properties were investigated. The dichroic ratio of 1.6 was calculated from the polarized EL spectra [Kawamoto et al., 2003]. To improve the polarized EL properties, a carbazole moiety as a hole-transporting unit in the same side chain was used to synthesize PM6-OXD-MCz showing a nematic LC phase. The fabricated single-layer EL device based on the aligned LCP exhibited good polarized EL properties, as shown in the right of Fig. 5.16.

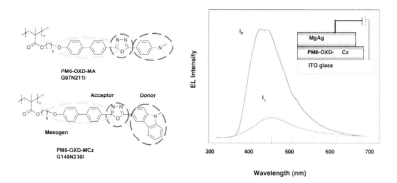

Figure 5.16 Molecular structures and LC phase behavior of LCPs for single-layer EL devices (left) and an example of polarized EL performance of single-layer devices fabricated with anisotropic PM6-OXD-Cz (right). $I_\parallel$ and $I_\perp$ are light intensity parallel and perpendicular to the alignment direction of LCPs.

5.5 Holographic Applications

5.5.1 Holographic Recording

Holography is a unique technique that enables concomitant recording of both phases and amplitudes of light waves. As shown in Fig. 5.17, the most fascinating feature of holography is that it can record and display a complete 3D image of an object [Bieringer, 2000; Kogelnik, 1969]. This provides unique opportunities for the next-generation storage technique by a simple recording and reading process.

In holography, the phase and amplitude of light waves are recorded by periodic alternation of physical properties of materials. According to the manner of recording of interference patterns, holograms are mainly classified into two types [Collier et al., 1971; Smith, 1977]. As shown in Fig. 5.18, one is an amplitude-type hologram, in which the interference pattern is recorded as a density variation in recording media. The other is a phase-type hologram, in which fringe patterns are recorded as a change in surface structure or refractive index. Theoretically, the diffraction efficiency of phase-type holograms is always higher than that of amplitude-type ones. Accordingly, the phase-type holograms for data recording are superior to the latter and most studies on holography are related to the phase-type ones.

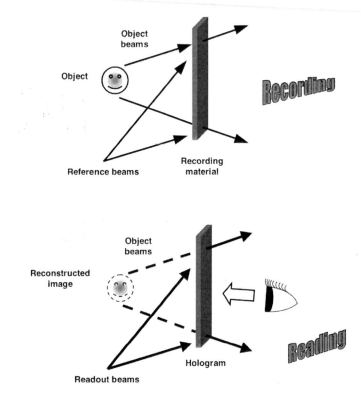

Figure 5.17 Schematic illustration of holographic data storage.

5.5.2 Recording Gratings with LC Alignment Changes

Azobenzene-containing LCPs with photoresponsive functions can be easily modulated into hierarchical patterns by adjusting the input light with wavelength, intensity, polarization, phase, interference, and so on. Therefore, holograms can be recorded in azobenzene-containing LC materials by inducing an orientation change of LC molecules in a periodic pattern obtained from interference of two coherent beams, the object and the reference beams. LC materials can bring about a larger change in refractive index comparing with amorphous ones. An alignment change of LC materials in a periodic fashion can be easily induced upon irradiation with interference patterns by overlapping two coherent beams, resulting in a large refractive-index modulation. This can contribute to a high diffraction efficiency of recorded gratings. In films of azobenzene-containing LCPs, both surface-relief and refractive-index gratings can be recorded correspondingly.

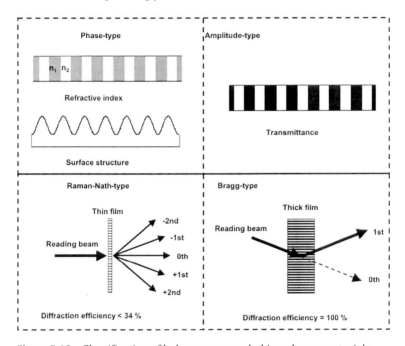

Figure 5.18 Classification of holograms recorded in polymer materials.

Wendorff et al. showed that holographic gratings could be inscribed in LC polymers composed of azobenzene moieties and photoinert mesogenic groups [Eich et al., 1987; Eich & Wendorff, 1987; Anderle & Wendorff, 1994]. Hvilsted et al. [1995] achieved holographic gratings with diffraction efficiencies of about 40% in azobenzene-containing LC polyesters. Recently, formation of holographic gratings in side-chain LCPs containing azobenzene moieties by means of photochemical phase transition was intensively studied. Hasegawa et al. [1999a & b] achieved dynamic holographic gratings by means of photochemical nematic-to-isotropic phase transition in LC copolymers containing an azobenzene moiety with strong donor–acceptor substituents in side chain (pseudostilbene-type azobenzenes). As shown in Fig 5.19, the inscription and removal of holographic gratings with a narrow fringe spacing of 1.4 µm was obtained within 150 and 190 ms, respectively. Moreover, the optical-switching behaviors of the holographic diffraction were observed repeatedly by turning the writing beams on and off. Using a siloxane group as a spacer in preparation of LC polymers decreased the glass transition temperature of the designed materials. This was because of the flexibility of the siloxane unit, which resulted in an effectively photoinduced nematic-to-isotropic phase transition at room temperature and formation of real-time holographic gratings [Hasegawa et al., 1999c].

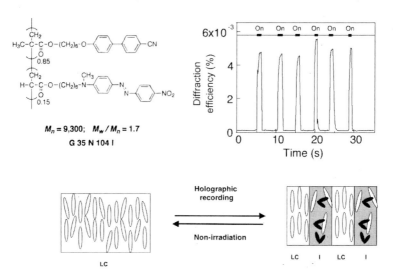

Figure 5.19 Dynamic holographic gratings by means of photochemical phase transition in films of azobenzene-containing LCPs.

5.5.3 Recording Gratings with Photoinduced Phase Transition

In a nematic LCP with azobenzenes as side mesogens, both surface-relief and refractive-index modulations were recorded, as shown in Fig. 5.20. In nematic LC phase, the obtained gratings exhibited a higher diffraction efficiency than that recorded in the glassy state even though the amplitude of surface-relief structures of the former was lower than that of the latter one [Yamamoto et al., 1999]. These indicated that the contribution of refractive-index gratings to the diffraction efficiency was larger than that of surface-relief gratings. Based on the same materials, holographic image storage using a photomask as an object was successfully obtained [Yamamoto et al., 2000]. Owing to spatial modulation of molecular alignment in the interference pattern, an alternating arrangement of LC and isotropic phases can be clearly observed with POM. It has been proved that such a system is capable of holographic recording of 3D objects with high resolution.

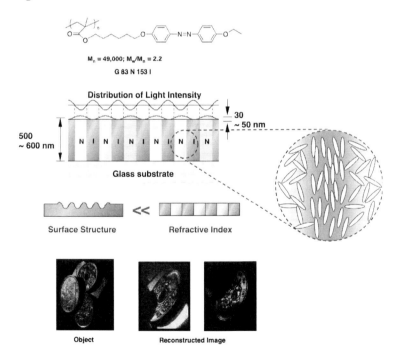

Figure 5.20 Holographic gratings recorded in photoresponsive LCPs with azobenzenes as nematic mesogens.

Using two phase-type gratings recorded in LC cells, grating waveguide couplers with a flat surface were fabricated as shown in Fig. 5.21 [Bang et al., 2007]. When a probe beam at 633 nm was incident to one grating, the beam propagated in the waveguide and an output beam came out from the other grating with the throughput coupling efficiency of about 5%. Upon irradiation of the LCP film between the two gratings with UV light to cause *trans–cis* photoisomerization and order–disorder transition of the azobenzene moiety, the intensity of the output beam was repeatedly switched upon alternating irradiation of visible light. It was found that the alternating irradiation at 366 and 436 nm induced reversible changes in the intensity of the output beam.

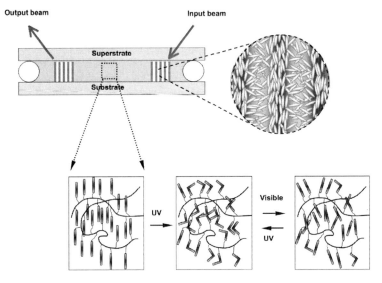

Figure 5.21 Waveguide couplers based on periodically flat-structured gratings and their photoswitching behaviors.

5.5.4 Subwavelength Gratings

Due to MCM of mesogens in LCPs, it is difficult to record holographic gratings with a narrow periodicity in a subwavelength scale [Yu et al., 2008]. Pretreatment with UV irradiation to induce *cis*-azobenzene-rich isotropic phase can eliminate such molecular cooperative effect of mesogens and enhance the photoresponse in a lowly viscous state. This enabled us to achieve the subwavelength modulation of surface relief and refractive index with interference patterns in Fig. 5.22 [Yu et al., 2008]. The surface relief of less than 10 nm and the refractive-index modulation were detected by atomic force microscopy (AFM) in tapping mode and phase mode, respectively. A large phase retardation and formed birefringence were observed in the recorded subwavelength gratings.

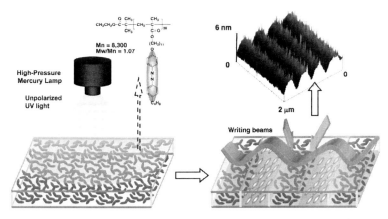

Figure 5.22 Subwavelength modulation of surface relief and refractive index in LCP films with pre-irradiation treatment.

5.5.5 Mechanically Tunable Gratings

Being one of commercially available products, an ABA-type triblock copolymer, polystyrene (PS)–b–polybutadiene–b–PS, is famous for its thermoplastics. The hard block of PS with a content of 20–30 wt.% forms the minority phase upon microphase separation, which acts as physical crosslinks for the majority phase of the soft block of rubbery polybutadiene (PB). Mechanical stretching can induce a large elastic deformation with recoverable properties. By applying this concept, block copolymers with thermoplastics were prepared [Bai & Zhao, 2001; 2002; Zhao et al., 2002]. Upon stretching-induced elastic deformation of grating samples recorded in the thermoplastic block copolymers, fringe spacing or grating periodicity was successfully adjusted, as shown in Fig. 5.23.

Recent progress in chemistry and material science enables one to freely design functional materials with suitable processes to satisfy the need of advanced materials for a variety of applications. Integration of azobenzene materials with other functionalized materials and processing provides the designed materials for holograms with special features. For instance, chemically crosslinking of azobenzene-containing polymer films with surface-relief gratings (SRGs) could fix the obtained surface modulation, producing permanent shape change [Zettsu et al., 2008].

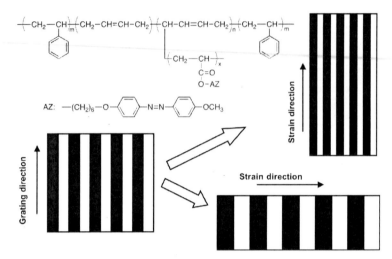

Figure 5.23 Mechanically tunable gratings recorded in azobenzene-containing LCPs showing thermoplastic properties.

5.5.6 Bragg-Type Gratings

As shown in Fig. 5.18, the Raman-Nath-type hologram recorded in a thin film exhibited a theoretically maximum diffraction efficiency of about 34%. This was far lower than the Bragg-type hologram in a thick film with a maximum diffraction efficiency of 100%. Furthermore, angular multiplicity could be easily obtained in the Bragg hologram, enabling thick films to be suitable for volume storage. To prepare thick and transparent films, amorphous and highly transparent PS blended with their block copolymers containing an azobenzene block as the minority phase to eliminate the scattering by utilizing the nanoscale microphase separation [Häckel et al., 2005; 2007]. Since all of the mesogens were confined in nanospheres dispersed in PS matrix, thick and optically transparent films were obtained by injection-molded method. These films showed good angle-multiplexing capability, in which 200 holograms were superimposed and reconstructed independently at the same spot and more than 1000 write/erase cycles were successfully obtained.

However, the density of photoresponsive mesogens in the block copolymer blend systems was very low, leading to a low diffraction efficiency of the recorded gratings. To induce a large change in refractive index and record Bragg gratings with high diffraction efficiency, a series of amorphous random copolymers with both azobenzene moieties and photoinert mesogens was prepared, as shown in Fig. 5.24 [Saishoji et al., 2007; Ishiguro et al., 2007]. The highly transparent poly(methyl methacrylate) (PMMA) was utilized as substrate, when $x > 50$ mol% in the materials preparation. Cyanobiphenyl and tolane groups were selected as the photoinert part to enhance the refractive-index modulation. Azobenzene mesogens with a low content (Z = about 5 mol%) acted as photoresponsive moieties. Transparent and thick films (>100 µm) were fabricated using a melting and pressing process.

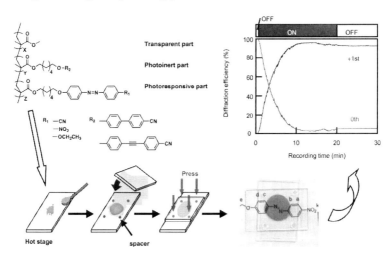

Figure 5.24 Preparation of transparent and thick films of azobenzene-containing copolymers for Bragg-type gratings.

Upon irradiation with the writing beams, the first-order diffracted beam (+1st) appeared immediately and the intensity of the zeroth-order beam (0th) decreased (Fig. 5.24). The maximum diffraction efficiency reached above 98% in all the prepared films. Furthermore, the polarization-selective multiple holographic data storage could be obtained using the photoinduced anisotropy as well as rewritable holographic recording with about 100% diffraction efficiency. These materials prepared by a simple but balanced formulation would provide a new guideline for the construction of high-performance holographic devices.

5.6 Other Applications

5.6.1 Photocontrol of Functional Materials

On the surface of oriented LCP films, not only LCs but also many functional materials can be aligned. These functional materials include dichroic dyes, electroluminescent materials, conducting polymer materials, as well as inorganic silica materials. It has been reported that mesoporous silica film can be deposited on the surface of photoaligned azobenzene-containing LCP films [Kawashima et al., 2002; 2004], as shown in Fig. 5.25. However, the orientation of azobenzene LCP film shows bad thermal stability (below 40°C); therefore, the structure of channels lacks mechanical stability due to insufficient crosslinking of the siloxane condensation. To overcome the above-mentioned problems, a new versatile photoalignment method using a photocrosslinkable cinnamate-containing LCPs was adopted [Fukumoto et al., 2005; 2006], in which the mesogenic biphenyl group is combined with a cinnamoyl terminal group.

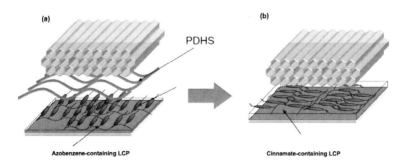

Figure 5.25 Photoalignment of silica mesochannels on films of photoresponsive LCPs based on (a) the photoisomerization of azobenzenes and (b) photocrosslinking cinnamates. Reprinted with permission from Fukumoto et al., Copyright 2005, WILEY-VCH Verlag GmbH & Co. KGaA, Weinheim.

Another advantage of using the photocrosslinkable LCP is that the resultant photochemical crosslinking firmly fixes the anisotropic molecular orientation. This feature is anticipated to allow facile performances of siloxane condensation at higher temperatures without damaging the alignment polymer film. Thus, the alignment stability was greatly enhanced.

5.6.2 Photocontrolled Nanostructures

Generally, photoresponsive LCPs show hydrophobic properties due to their carbon-rich molecular structures. If one photoresponsive LCP is designed as one block and one hydrophilic polymer as another block, amphiphilic LC diblock copolymers (LCBCs) can be obtained with photocontrollable microphase separated structures. These will be discussed in Chapter 7. Recently, a novel type of LCPs with a block mesogenic side-group was proposed as shown in Fig. 5.26 [Okano et al., 2009]. The block mesogenic molecule contains a hydrophobic azotolane moiety and two hydrophilic oligooxyethylene groups in the same unit, bringing about the designed LCP with self-organization, amphiphilic, and photoresponsive properties. Similarly to photoresponsive LC block copolymers, the microphase-separated lamellar nanostructures can be reversibly controlled upon irradiation of LPL or unpolarized-UV light.

5.6.3 Photorewritable Paper

Multi-bilayered films have been regarded as 1D photonic band-gap materials. If one layer is photoresponsive LCP material, the reflectivity of the multi-bilayered materials should be photocontrolled as shown in Fig. 5.27 [Moritsugu et al., 2011]. Kurihara et al. [1991] reported that the reflectance of the multi-bilayered film with azobenzene LCP as one layer was increased by UV irradiation because of the transformation from the out-of-plane orientation to an in-plane random orientation of azobenzene mesogens. By this way, on–off switching of the reflection is achieved by alternating the thermally spontaneous out-of-plane orientation and the following photoisomerization (simultaneously smectic LC to isotropic phase transition) of LCPs. Such a multi-bilayered film has promised its application as photo-rewritable paper materials.

162 | Liquid Crystal Polymers

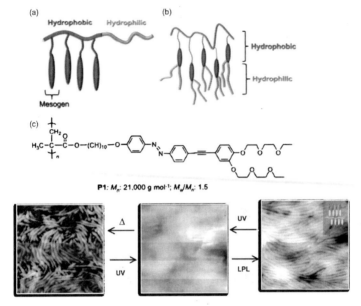

Figure 5.26 Photocontrol microphase-separated nanostructures of LCPs with a block mesogenic side-group. Reprinted with permission from Okano et al., Copyright 2009, WILEY-VCH Verlag GmbH & Co. KGaA, Weinheim.

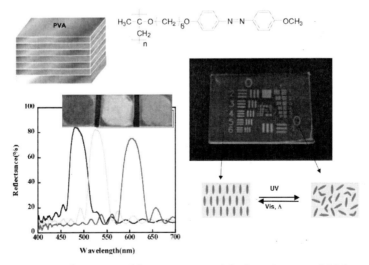

Figure 5.27 Photorewritable paper materials based on multi-bilayers containing photoresponsive LCPs. Reprinted with permission from Moritsugu et al., Copyright 2011, WILEY-VCH Verlag GmbH & Co. KGaA, Weinheim.

5.6.4 Photoswitching of Gas Permeation

When photoisomerization and photoinduced phase transition occurs in photoresponsive LCP films, there are not additional changes to the chemical composition upon light irradiation. However, this photoresponse can cause interesting changes in materials. For instance, photocontrol of gas permeation through films of one photoresponsive LCP was obtained as shown in Fig. 5.28. This photoswitching behavior was carried out by photocontrolled LC orientation between a mono-domain orientational ordered phase and an isotropic LC glass upon irradiation with circularly polarized light and non-polarized blue light [Liu et al., 2011].

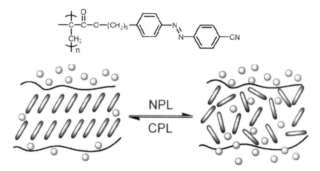

Figure 5.28 Photoswitching of gas permeation through films of one photoresponsive LCP. Reprinted with permission from Liu et al., Copyright 2011, WILEY-VCH Verlag GmbH & Co. KGaA, Weinheim.

5.6.5 Photodriven Motions

By carefully molecular design, the photoisomerization-induced change in molecular shape and the accompanied phase transition can be amplified into macroscopic scale. As shown in Fig. 5.29, a brush-like LCP made up of a polymethacrylate backbone with outstretched side-chains azobenzene mesogens can be assembled into free-standing films [Hosono et al., 2010]. With the help of stretching with Teflon sheets, the film with 3D macroscopic ordering can literally bend and stretch upon alternating irradiation by UV and visible light. This directly converts light energy directly into a mechanical force. However, the photomechanical film should be processed in the LC phase at higher temperature in order for the 3D order.

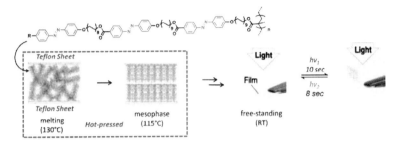

Figure 5.29 Photomechanical films of brush-like photoresponsive LCPs.

To develop room-temperature photomechanical system, a hybrid LC films in Fig. 5.30 was prepared with light-responsive LCP microparticles showing wrinkled morphologies as guest and photoinert polymers as host [Yu et al., 2011]. PM6AZOC2 was chosen because a nematic mesogen often shows a low viscosity and quicker photoresponse than a smectic one. LCP microparticles with wrinkled morphologies were fabricated with a phase reversion method in a THF–ethanol mixed solution. This unique feature of interesting surface morphologies can increase the specific surface area of the LCP microparticles comparing with smooth ones, which could help improvement of their photoresponse.

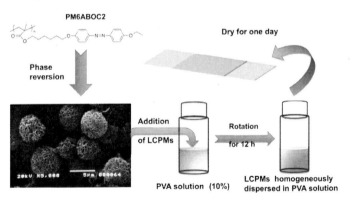

Figure 5.30 Fabrication of hybrid film with photoresponsive LCP microparticles. Reprinted with permission from Liu et al., Copyright 2011, WILEY-VCH Verlag GmbH & Co. KGaA, Weinheim.

Then, the hybrid LC film was mechanically stretched at a certain elongation rate with a tensile test machine. After this mechanical treatment, a structural anisotropy was clearly observed as a result of shape deformation of the LCP microparticles from spheres to ellipsoids. Furthermore, the mesogens in the LCP microparticles were homogenously aligned along the stretching direction (Fig. 5.31).

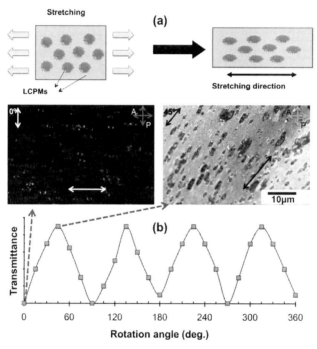

Figure 5.31 Schematic illustration of the mechanically stretched process of the hybrid LC films. (a) Possible scheme of LCPMs in the hybrid films. (b) Polarizing optical microscopic images of the mechanically stretched hybrid LC films and transmittance through two crossed polarizers as a function of rotation angle in relation to the stretching direction (A, analyzer; P, polarizer). Reprinted with permission from Yu et al., Copyright 2011, WILEY-VCH Verlag GmbH & Co. KGaA, Weinheim.

The photoresponsive behaviors of the stretched hybrid LC films with a strong anisotropy were studied using an unpolarized UV light beam at 360 nm. Upon irradiation, the hybrid film slowly bent toward the light source, just along the stretching direction. A 10 s photoirradiation induced a bending angle of about 10°. Then 20 s later, the bending angle of the hybrid film increased to 45°. A bending angle of about 90° was obtained after the film was irradiated for

40 s (Fig. 5.32). Then, a much longer photoirradiation almost did not cause change the bent film. Turning off the actinic light, the photoinduced deformation remained without detectable relaxation, indicating that the bent films were obtained in a photostationary state. In fabrication of the hybrid films, the mechanical stretching played a key role in the integrating mechanical and photoresponse in the hybrid LC films with photoinduced macroscopic deformation. Without this treatment, mechanical and photoresponse cannot be coupled efficiently.

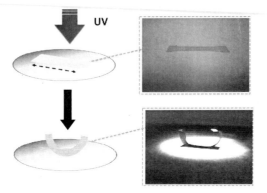

Figure 5.32 Photoresponsive behaviors of hybrid film hybrid film upon mechanical stretching. Reprinted with permission from Yu et al., Copyright 2011, WILEY-VCH Verlag GmbH & Co. KGaA, Weinheim.

In the hybrid films, the confinement effect by the PVA matrix might provide the LCP microparticles with additionally elastic properties, which is difficulty to be obtained in the side-chain type LCP films. Besides, the photomechanical behaviors were ascribed to a bimetal model shown in Fig. 5.33, similarly to that of 3D crosslinked LC elastomers (LCEs) which will be discussed in Chapter 6. According to the fabrication process of the hybrid film, the LCP microparticles were homogeneously dispersed in the mixed solution, which was then cast on glass slides. Upon slow evaporation of the solvent, the microparticles gradually floated to the upper of the solution because of the lower density of the PM6ABOC2 than water. Therefore, the LCP microparticles were inclined to distribute in the upper layer of the hybrid film, and the lower layer of the film showed a lower dispersity of the microparticles (Fig. 5.33).

Other Applications | 167

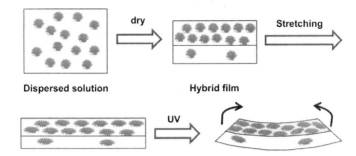

Figure 5.33 Possible bi-layered structures of the fabricated hybrid film hybrid films. Reprinted with permission from Yu et al., Copyright 2011, WILEY-VCH Verlag GmbH & Co. KGaA, Weinheim.

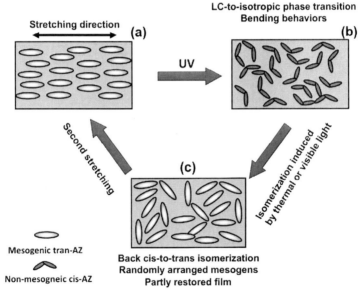

Figure 5.34 Possible schemes of LC alignment in the LCP microparicles of the hybrid LC films. (a) The LCs were aligned along the stretching direction upon stretching. (b) Upon UV irradiation, photoisomerization and LC-to-isotropic phase transition occurred, inducing a bending behavior. (c) Thermal or visible light induced back-isomerization of azobenzenes, but the ordered LC phase was not restored because of the restriction of the polymer chain below its T_g. Reprinted with permission from Yu et al., Copyright 2011, WILEY-VCH Verlag GmbH & Co. KGaA, Weinheim.

Reversibility is one of the most fascinating features of azobenzene-containing photoresponsive LCP materials. But in the present hybrid LC system, the bent film was only partly restored and its original flat state was not completely recovered upon irradiation of visible light. It has been reported that the LC phase of the LCPs cannot be obtained at room temperature even though *cis*-to-*trans* back isomerization of azobenzene moieties occurs (Fig. 5.34c) because of the restriction of polymer chain below its glass transition temperature (T_g). Therefore, the ordered state already induced by the mechanical stretching could not be directly achieved (from b to a as shown in Fig. 5.34). However, the random mesogens can be re-aligned by re-stretching the contracted films (from c to a as shown in Fig. 5.34), endowing the secondly stretched film with a similar photomechanical response. Thus, the photomechanical performances of the hybrid film were accomplished in a reversible way.

References

Anderle, K. and Wendorff, J. H. (1994). Holographic recording using liquid crystalline side chain polymers, *Mol. Cryst. Liq. Cryst.* **243**, pp. 51–75.

Bai, S. and Zhao, Y. (2001). Azobenzene-containing thermoplastic elastomers: Coupling mechanical and optical effects, *Macromolecules* **34**, pp. 9032–9038.

Bai, S. and Zhao, Y. (2002). Azobenzene elastomers for mechanically tunable diffraction gratings, *Macromolecules* **35**, pp. 9657–9664.

Bang, C. U., Shishido, A., and Ikeda, T. (2007). Azobenzene liquid-crystalline polymer for optical switching of grating waveguide couplers with a flat surface, *Macromol. Rapid Commun.* **28**, pp. 1040–1044.

Bieringer, T. (2000). Holographic data storage. In: Coufal, H. J., Psaltis, D., Sincerbox, G. T., editors. *Springer Series in Optical Sciences*. New York: Springer, p. 209.

Collier, R. J., Burckhardt, C. B., and Lin, L. H. (1971). *Optical holography*. Colier, R., editor. New York: Academic Press.

Eich, M., Wendorff, J. H., Reck, B., and Ringsdorf, H. (1987). Reversible digital and holographic optical storage in polymeric liquid crystals, *Makromol. Chem. Rapid Commun.* **8**, pp. 59–63.

Eich, M. and Wendorff, J. H. (1987). Erasable holograms in polymeric liquid crystals, *Makromol. Chem. Rapid Commun.* **8**, pp. 467–471.

Fukumoto, H., Nagano, S., Kawatsuki, N., and Seki, T. (2005). Photo-orientation of mesoporous silica thin films on photo-crosslinkable polymer film, *Adv. Mater.* **17**, pp. 1035–1039.

Fukumoto, H., Nagano, S., Kawatsuki, N., and Seki, T. (2006). Photo-alignment behavior of mesoporous slica thin films synthesized on a photo-crosslinkable polymer films, *Chem. Mater.* **18**, pp. 1226–1234.

Grell, M. and Bradley, D. D. C. (1999). Polarized luminescence from oriented molecular materials, *Adv. Mater.* **11**, pp. 895–905.

Häckel, M., Kador, L., Kropp, D., Frenz, C., and Schmidt, H. (2005). Holographic gratings in diblock copolymers with azobenzene and mesogenic side groups in the photoaddressable dispersed phase, *Adv. Funct. Mater.* **15**, pp. 1722–1727.

Häckel, M., Kador, L., Kropp, D., and Schmidt, H. (2007). Polymer blends with azobenzene-containing block copolymers as stable rewritable volume holographic media, *Adv. Mater.* **19**, pp. 227–231.

Hasegawa, M., Yamamoto, T., Kanazawa, A., Shiono, T., and Ikeda, T. (1999a). Dynamic grating using photochemical phase transition of polymer liquid crystals containing azobenzene derivatives, *Adv. Mater.* **11**, pp. 675–677.

Hasegawa, M., Yamamoto, T., Kanazawa, A., Shiono, T., and Ikeda, T. (1999b). Photochemically-induced dynamic grating by means of side-chain polymer liquid crystals, *Chem. Mater.* **11**, pp. 2764–2769.

Hasegawa, M., Yamamoto, T., Kanazawa, A., Shiono, T., Ikeda, T., Nagase, Y., Akiyama, E., and Takamura, Y. (1999c). Real-time holographic grating by means of photoresponsive polymer liquid crystals with flexible siloxane spacer in the side chain, *J. Mater. Chem.* **9**, pp. 2765–2772.

Hosono, N., Kajitani, T., Fukushima, T., Ito, K., Sasaki, S., Takata, M., and Aida, T. (2010). Large-area three-dimensional molecular ordering of a polymer brush by one-step processing, *Science* **330**, pp. 808–811.

Hvilsted, S., Andruzzi, F., Kulinna, C., Siesler, H. W., and Ramanujam, P. S. (1995). Novel side-chain liquid crystalline polyester architecture for reversible optical storage, *Macromolecules* **28**, pp. 2172–2183.

Ikeda, T., Horiuchi, S., Karanjit, D. B., Kurihara, S., and Tazuke, S. (1988). Photochemical image storage in polymer liquid crystals, *Chem. Lett.*, pp. 1679–1682.

Ikeda, T., Kurihara, S., Karanjit, D. B., and Tazuke, S. (1990). Photochemically induced isothermal phase transition in polymer liquid crystals with mesogenic cyanobiphenyl side chains, *Macromolecules* **23**, pp. 3938–3943.

Ikeda, T. and Tsutsumi, O. (1995). Optocal switching and image storage by means of azobenzene liquid-crystal films, *Science* **268**, pp. 1873–1875.

Ishiguro, M., Sato, D., Shishido, A., and Ikeda, T. (2007). Bragg-type polarization gratings formed in thick polymer films containing azobenzene and tolane moieties, *Langmuir* **23**, pp. 332–338.

Kanazawa, A., Hirano, S., Shishido, A., Hasegawa, M., Tsutsumi, O., Shiono, T. Ikeda, T., Nagase, Y., Akiyama, E., and Takamura, Y. (1997). Photochemical phase transition behaviour of polymer azobenzene liquid crystals with exible siloxane units as a side-chain spacer, *Liq. Cryst.* **23**, pp. 293–298.

Kawamoto, M., Mochizuki, H., Shishido, A., Tsutsumi, O., Ikeda, T., Lee, B., and Shirota, Y. (2003). Side-chain polymer liquid crystals containing oxadiazole and amine moieties with carrier-transporting abilities for single-layer light-Eeitting diodes, *J. Phys. Chem. B* **107**, pp. 4887–4893.

Kawashima, Y., Nakagawa, M., Seki, T., and Ichimura, K. (2002). Photo-orientation of mesostructured silica via hierarchical multiple transfer, *Chem. Mater.* **14**, pp. 2842–2844.

Kawashima, Y., Nakagawa, M., Ichimura, K., and Seki, T. (2004). Photoorientation of mesoporous silica matcrials via transfer from an azobenzene-containing polymer monolayer, *J. Mater. Chem.* **14**, pp. 328–335.

Kawatsuki, N., Yamashita, A., Fujii, Y., Kitamura, C., and Yoneda, A. (2008). Thermally enhanced photoinduced reorientation in photo-crosslinkable liquid crystalline polymers comprised of cinnamate and tolane mesogenic groups, *Macromolecules* **41**, pp. 9715–9721.

Kogelnik, H. (1969). Coupled wave theory for thick hologram gratings, *Bell Syst. Technol. J.* **48**, pp. 2909–2946.

Liu, J., Wang, M., Dong, M., Gao, L., and Tian, J. (2011). Reversible photoinduced switching of permeability in a cast non-porous film comprising azobenzene liquid crystalline polymer, *Macromol. Rapid Commun.* **32**, pp. 1557–1562.

Moritsugu, M., Ishikawa, T., Kawata, T., Ogata, T., Kuwahara, Y., and Kurihara, S. (2011). Thermal and photochemical control of molecular orientation of azo-functionalized polymer liquid crystals and application for photo-rewritable paper, *Macromol. Rapid Commun.* **32**, pp. 1546–1550.

Mochizuki, H., T. Hasui, T. Shiono, T. Ikeda, C. Adachi, Y. Taniguchi, and Shirota, Y. (2000a). Emission behavior of molecularly-doped electroluminescent device using liquid-crystalline matrix, *Appl. Phys. Lett.* **77**, pp. 1587–1589.

Mochizuki, H., Hasui, T., Kawamoto, M., Shiono, T., Ikeda, T., Adachi, C., Taniguchi, Y., and Shirota, Y. (2000b). Novel liquid-crystalline and amorphous materials containing oxadiazole and amine moieties for electroluminescent devices, *Chem. Commun.* pp. 1923–1924.

Mochizuki, H., Hasui, T., Kawamoto, M., Ikeda, T., Adachi, C., Taniguchi, Y., and Shirota, Y. (2003). A novel class of photo- and electroactive polymers containing oxadiazole and amine moieties in a side chain, *Macromolecules* **36**, pp. 3457–3464.

Okano, K., Shishido, A., and Ikeda, T. (2006a). An azotolane liquid-crystalline polymer exhibiting extremely large birefringence and its photoresponsive behavior, *Adv. Mater.* **18**, pp. 523–527.

Okano, K., Tsutsumi, O., Shishido, A., and Ikeda, T. (2006b). Azotolane liquid-crystalline polymers: Huge change in birefringence by photoinduced alignment change, *J. Am. Chem. Soc.* **128**, pp. 15368–15369.

Okano, K., Shishido, A., and Ikeda, T. (2006c). Photochemical phase transition behavior of highly birefringent azo-tolane liquid-crystalline polymer films: Effects of the position of the tolane group and the donor-acceptor substituent in the mesogen, *Macromolecules* **39**, pp. 145–152.

Okano, K., Mikami, Y., Shibata, Y., and Yamashita, T. (2008). Develoment of novel azobenzene derivatives showing a room temperature liquid-crystalline phase and their photoresponsive behavior, *J. Photopoly. Sci. Tech.* **21**, pp. 549–552.

Okano, K., Mikami, Y., and Yamashita, T. (2009). Liquid-crystalline polymer with a block mesogenic side group: Photoinduced manipulation of nanophase-separated structures, *Adv. Funct. Mater.* **19**, pp. 3804–3808.

Peng, Z., Bao, Z., and Galvin, M. E. (1998). Polymers with bipolar carrier transport abilities for light emitting diodes, *Chem. Mater.* **10**, pp. 2086–2090.

Saishoji, A., Sato, D., Shishido, A., and Ikeda, T. (2007). Formation of Bragg gratings with large angular multiplicity by means of the photoinduced reorientation of azobenzene copolymers, *Langmuir* **23**, pp. 320–326.

Sekine, C., Iwakura, K., Konya, N., Minami, M., and Fujisawa, K. (2001). Synthesis and properties of high birefringence liquid crystals: Thiophenylacetylene and benzothiazolylacetylene derivatives, *Liq. Cryst.* **28**, pp. 1361–1367.

Shishido, A., Tsutsumi, O., Kanazawa, A., Shiono, T., Ikeda, T., and Tamai, N. (1997). Rapid optical switching by means of photoinduced change in refractive index of azobenzene liquid crystals detected by reflection-mode analysis, *J. Am. Chem. Soc.* **119**, pp. 7791-7796.

Smith, H. M. (1977). *Holographic recording materials.* H. M. Smith, editor. Springer-Verlag: Berlin.

Tsutsumi, O., Kitsunai, T., Kanazawa, A., Shiono, T., and Ikeda, T. (1998). Photochemical phase transition behavior of polymer azobenzene liquid crystals with electron-donating and -accepting substituents at the 4,4'-positions, *Macromolecules* **31**, pp. 355-359.

Uekusa, T., Nagano, S., and Seki, T. (2009). Highly ordered in-plane photoalignment attained by the brush architecture of liquid crystalline azobenzene polymer, *Macromolecules* **42**, pp. 312-318.

Wu, Y., Demachi, Y., Tsutsumi, O., Kanazawa, A., Shiono, T., and Ikeda, T. (1998a). Photoinduced alignment of polymer liquid crystals containing azobenzene moieties in the side chain. 1. effect of light intensity on alignment behavior, *Macromolecules* **31**, pp. 349-354.

Wu, Y., Demachi, Y., Tsutsumi, O., Kanazawa, A., Shiono, T., and Ikeda, T. (1998b). Photoinduced alignment of polymer liquid crystals containing azobenzene moieties in the side chain. 2. effect of spacer length of the azobenzene unit on alignment behavior, *Macromolecules* **31**, pp. 1104-1108.

Wu, Y., Demachi, Y., Tsutsumi, O., Kanazawa, A., Shiono, T., and Ikeda, T. (1998c). Photoinduced alignment of polymer liquid crystals containing azobenzene moieties in the side chain. 3. effect of structure of photochromic moieties on alignment behavior, *Macromolecules* **31**, pp. 4457-4463.

Yamamoto, T., Hasegawa, M., Kanazawa, A., Shiono, T., and Ikeda, T. (1999). Phase-type gratings formed by photochemical phase transition of polymer azobenzene liquid crystals: Enhancement of diffraction efficiency by spatial modulation of molecular alignment, *J. Phys. Chem. B* **103**, pp. 9873-9878.

Yamamoto, T., Hasegawa, M., Kanazawa, A., Shiono, T., and Ikeda, T. (2000). Holographic gratings and holographic image storage by photochemical phase transition of polymer azobenzene liquid-crystal films, *J. Mater. Chem.* **10**, pp. 337-342.

Yoneyama, S., Yamamoto, T., Tsutsumi, O., Kanazawa, A., Shiono, T., and Ikeda, T. (2002). High-performance material for holographic gratings by means of a photoresponsive polymer liquid crystal containing a tolane moiety with high birefringence, *Macromolecules* **35**, pp. 8751-8758.

Yu, H. F., Shishido, A., and Ikeda, T. (2008). Subwavelength modulation of surface relief and refractive index in pre-irradiated liquid-crystalline polymer films, *Appl. Phys. Lett.* **92**, pp. 103117 (1–3).

Yu, H. F., Kobayasi, T., and Ge, Z. (2009). Precise control of photoinduced birefringence in azobenzene-containing liquid-crystalline polymers by post functionalization, *Macromol. Rapid Commun.* **30**, pp. 1725–1730.

Yu, H., Dong, C., Zhou, W., Kobayashi, T., and Yang, H. (2011). Wrinkled liquid-crystalline microparticle-enhanced photoresponse of PDLC-like films by coupling with mechanical stretching, *Small* **7**, pp. 3039–3045.

Zettsu, N., Ogasawara, T., Mizoshita, N., Nagano, S., and Seki, T. (2008). Photo-triggered surface relief grating formation in supramolecular liquid crystalline polymer systems with detachable azobenzene units, *Adv. Mater.* **20**, pp. 516–521.

Zhao, Y., Bai, S., Dumont, D., and Galstian, T. (2002). Mechanically tunable diffraction gratings recorded on an azobenzene elastomer, *Adv. Mater.* **14**, pp. 512–514.

Chapter 6

Liquid Crystal Elastomers

From low-molecular-weight (LMW) compounds, high-molecular-weight polymers (macromolecules) to crosslinked elastomers, the interaction among photoresponsive mesogens in systems gradually increases. In addition, the stability, movability, and photoresponse in the three systems are far different as shown in Fig. 6.1.

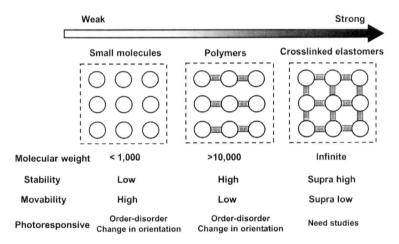

Figure 6.1 Description of photoresponsive mesogens in three systems.

LC elastomers (LCEs) can be prepared by directly crosslinking conventional LCPs into a network, which integrates mechanical

---

Dancing with Light: Advances in Photofunctional Liquid-Crystalline Materials
Haifeng Yu
Copyright © 2015 Pan Stanford Publishing Pte. Ltd.
ISBN 978-981-4411-11-0 (Hardcover), 978-981-4411-12-7 (eBook)
www.panstanford.com

properties of elastomers with regular ordering of LC materials by the network topologies, as shown in Fig. 6.2. In three-dimensional (3D) crosslinked LCEs, the initial ordering of mesogens can be fixed by the crosslinkers, which might give rise to quick change in shape due to a fast order–disorder transition, induced by slight changes (light, thermal, and so on) in the orientational order of mesogens [de Gennes, 1975].

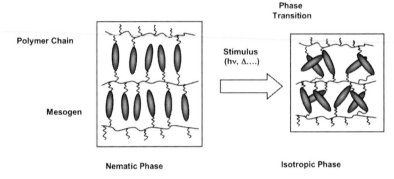

Figure 6.2 Schematic illustration of responsive LCEs.

Combination of the anisotropic aspects of LC phases and the rubber elasticity of polymer networks enables LCEs to show unique features like anisotropic change in shape. A large deformation can be induced in response to external stimuli such as temperature, electricity, pH, or humidity. When a chromophore (e.g., an azobenzene) as a mesogen is introduced into LCEs, shape, and volume change in response to light can be produced, which can directly convert light energy into mechanical power. Thus, photomechanical and photomobile properties can be obtained by the photochemical phase transition and photoinduced alignment.

6.1 Preparation of LCEs

In general, LC materials are immiscible with linear polymers or macromolecular networks due to entropic reasons. Thus, it is difficult to prepare LCEs by swelling a traditional rubber with LMWLCs. Crosslinking of linear LCPs is one of the best ways to obtain freestanding LCE films. Here, several methods developed for preparing LCEs will be discussed in this section.

6.1.1 Two-Step Method

A Germany research group led by Finkelmann first successfully prepared LCEs by a two-step method, as shown in Fig. 6.3 [Finkelmann et al., 1981]. In this strategy, they elegantly utilized different reactive activity between vinyl groups and mathacryloyl groups when they react with polyhydrosiloxane. In the first step, owing to the fast reaction speed of vinyl groups, well-defined weak networks are synthesized. These networks are then deformed with a constant load to induce LC alignment in the weakly crosslinked networks. In the second step, the anisotropic network is further crosslinked to fix the ordered mesogens. The advantage of this method is that the induced network anisotropy in the first step is reproducible, so that well-aligned elastomers are fabricated. However, it is difficult for the obtained network to be purified and sometimes impurities are involved. This undoubtedly influences the photoresponsive performance of the obtained LCEs.

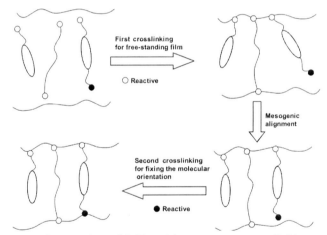

Figure 6.3 Preparation of LCEs with a two-step crosslinking method proposed by Finkelmann group in Germany.

6.1.2 One-Step Method of Direct Crosslinking of Linear LCPs

To get rid of the impurities in preparation of LCEs, the one-step method to prepare highly oriented LCEs was developed [Broer et al., 1989]. As shown in Fig. 6.4, linear LCPs containing additional functional groups for crosslinking should be first prepared, then in situ photopolymerization of macroscopically aligned LCPs was carried out. This method has the advantage that the LCPs can be purified and characterized before the crosslinking step. However, if not all the functional groups are crosslinked, linearly LCPs in an alignment state will remain in the resulted LCEs. Recently, an elegant method was used to prepare LCEs with a linear nematic LCPs by direct crosslinking upon electric-beam irradiation [Naka et al., 2011]. Thus, almost all the LCPs can be changed into LCEs by this simple method.

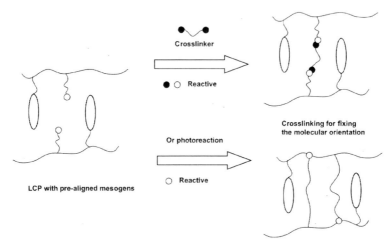

Figure 6.4 Preparation of LCEs with one-step method by post-crosslinking of pre-aligned linear LCPs with high purity.

6.1.3 Polymerization of LC Mixture of Monomers and Crosslinkers

As shown in Fig. 6.5, the third method is the direct polymerization of a mixture of LC monomers and crosslinkers under orientational condition of external fields. This mixture can be initiated by thermal or UV light to form anisotropic LCEs. Obviously, the orientation of LMWLC compounds is much easier than that of LCPs. At this case, it is not necessary for the crosslinkers to show an LC phase. For example, a homogeneously aligned LCE with photomechanical properties was prepared by thermal polymerization or photopolymerization.

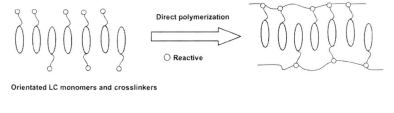

Figure 6.5 Preparation of LCEs by direct polymerization of LC mixtures of monomers and crosslinkers.

6.1.4 Physical Crosslinking Method

In addition to chemical crosslinking, physical crosslinking can be applied to create ordered LCEs. One of the typical examples is ABA triblock-copolymers, in which A is the hard-block (high T_g or crystalline material) and B is the LC "soft" block. The materials phase separates into a micellar structure thus fixing the LC block by glassy frozen spheres in a 3D network. Another example is based on self-assembly of block copolymers. As shown in Fig. 6.6, muscle-like materials with a lamellar structure based on a nematic triblock copolymer can phase separate into LCEs [Li et al., 2004]. The material consists of a repeated series of nematic (N) polymer blocks and conventional rubber (R) blocks. The synthesis of block copolymers with well-defined structures and narrow molecular-weight distributions is a crucial step in the creation of artificial muscle based on triblock elastomers.

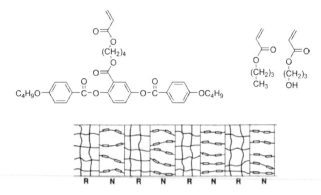

Figure 6.6 Preparation of LCEs with physical crosslinking (N, nematic; R, rubber).

6.2 Photochemical Phase Transition

LCEs combine characteristics of the two systems, rubbery elastomers with 3D network and self-organized LCPs with LC ordering, which have been regarded as a novel class of soft materials. In 1975, de Gennes first proposed the concept of LCEs as artificial muscles by taking advantage of their substantial uniaxial contraction in the direction of the director axis. Then, he theoretically predicted the possibility of a large deformation of LCEs induced by phase transition. Finkelmann et al. [1981] did pioneered work in preparation of LCEs with mesogens in alignment of polydomain and monodomain.

Usually, the crosslinking density has a great influence on the macroscopic properties and the phase structures of LCEs with crosslinked networks. On the one hand, the mobility of mesogenic segments is reduced with an increase in density of crosslinking points, and consequently, the mobility of mesogens are suppressed. On the other hand, such confinement from crosslinking may bring about larger change than a linear LCP in response to external stimuli. In fact, a linear LCP can be regarded as a special LCE with a crosslinking density of zero.

As previously stated in the textbook, the photoresponsive mesogens in LMW LCs and LCPs undergo photoinduced phase transition upon UV irradiation to induced *trans–cis* isomerization.

Similarly, azobenzenes in LCEs show analog photoresponse. Kurihara et al. [1998] prepared crosslinked LCEs containing azobenzene molecules through polymerization of ternary mixtures of monofunctional, di-functional LC monomers and a LMW azobenzene LC, and then evaluated their photoresponsive behavior. They demonstrated that the response time and the decay time were 1 μs and 100 μs, respectively. Such fast photoresponse can be ascribed to the suppression and confinement of motion of mesogenic groups by crosslinking, whereas non-crosslinked LCP analogs did not show such quick response. It has been concluded through intensive studies that the stabilization of an initial ordered state by way of crosslinking gives rise to a quick order–disorder transition, induced by slight changes in the orientational order of mesogens, and a fast disorder–order transition, due to relaxation of the strain generated upon photoisomerization of the azobenzene molecules [Kurihara et al., 1999].

6.3 Photoinduced Contraction and Expansion

In 1967, Lovrien first proposed the concept of photoviscosity effect. He pointed out that if the macromolecule changed dimensions under elastic constraint, there might occur a conversion of light energy to mechanical energy. However, the photoinduced deformation was too small when no phase transition occurred in amorphous photoresponsive elastomers. Such photomechanical effect was greatly enhanced in LCE systems.

Confining mesogens in 3D crosslinked networks, LCEs usually show thermoelastic properties: they contract along the alignment direction of mesogens upon thermally induced LC-isotropic phase transition, and expand upon cooling below the phase transition temperature. If photoresponsive mesogens such as azobenzene and cumarin are incorporated into LCEs, photoinduced motions such as contraction and expansion are expected by the above-mentioned photochemical phase transition (or photochemically induced decrease in LC ordering).

As shown in Fig. 6.7, an azobenzene molecule exhibits a large change in molecular configuration upon photoisomerization, and the distance between the 4- and 4'-carbons in benzene rings contracts from 9.0 Å of *trans*-azobenzene to 5.5 Å of *cis*-azobenzene. Such a photoinduced molecular change results in a large contraction ratio, which arouses intensive interest of scientists. Eisenbach first reported that amorphous polymers crosslinked with azobenzenes contracted upon UV irradiation to induce *trans–cis* isomerization, while it expanded by irradiation with visible light, which caused *cis–trans* back-isomerization [Eisenbach, 1980]. However, the observed contraction was only 0.15–0.25%, which is too small from the viewpoint of practical applications.

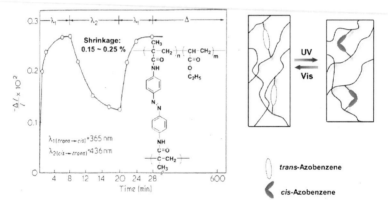

Figure 6.7 Photoinduced shrinkage by photoisomerization of non-LC materials.

Different from amorphous elastomers, photoinduced deformation has been greatly enhanced in LCEs. In 2001, Finkelmann et al. succeeded in inducing a contraction ratio of 20% in LCEs with polysiloxane main chains and azobenzene chromophores at the crosslinks upon UV exposure to cause the *trans–cis* isomerization of azobenzenes. After irradiation was stopped, the elastomers thermally returned to the original state due to *cis–trans* back-isomerization of the azobenzene mesogens (Fig. 6.8). As far as photomechanical effects are concerned, the subtle variation in nematic LC order upon *trans–cis* isomerization causes a large uniaxial deformation along the director axis when the LC molecules are strongly associated by covalent crosslinking to form a 3D polymer network.

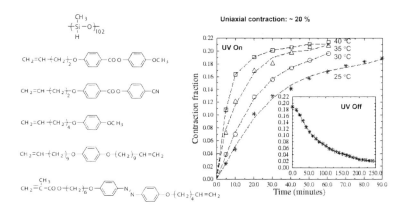

Figure 6.8 Photoinduced contraction in LCEs prepared with polysiloxane as matrix.

Hogan et al. [2002] incorporated a wide range of azobenzene derivatives into LCEs as photoresponsive moieties and examined their deformation behavior upon UV exposure. Li et al. [2003] synthesized monodomain nematic side-on LCEs containing azobenzenes by photopolymerization of aligned azobenzene monomers in conventional LC cells. Thin films of these LCEs showed fast (less than 1 min) photochemical contraction of up to 18% upon irradiation with UV light and a slow thermal recovering in the dark, as shown in Fig. 6.9. Then, they studied electrically induced contraction with a lamellar-structured triblock copolymer LCEs prepared by

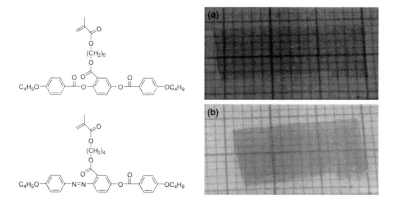

Figure 6.9 Photoinduced contraction in azobenzene side-on LCEs.

photopolymerization of aligned mesogen under a magnetic field [Li et al., 2004]. But they did not report the photomechanical effect since no photoresponsive mesogens are involved in the block copolymer-based LCEs.

6.3 Photoinduced 3D Motions

Although LCEs show excellent 2D motional properties of contraction and expansion, photoinduced 3D motions with diverse of ways such as bending, twisting, and rotation, are expected from the viewpoint of practical applications (Fig. 6.10). As described in Chapter 5, LCPs with a photoresponsive moiety in each mesogen show a faster photoresponse than LC copolymers with a low content of photochromic mesogens, so does in LCEs. It is expected that photomechanical and photomobile performances with a quick response to actinic light can be obtained in LCEs composed of only azobenzenes as photoresponsive mesogens [Ikeda et al., 2007].

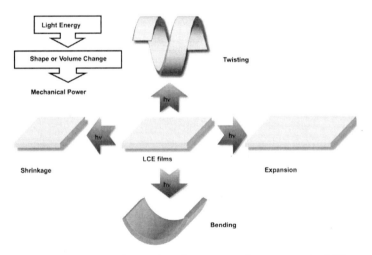

Figure 6.10 Diverse of photoinduced motion in photoresponsive LCEs.

6.3.1 Photoinduced Bending

As shown in Fig. 6.11, the photoresponsive behavior of monodomain aligned LCE films in both organic solvent and air was first reported in 2003 [Ikeda et al., 2003]. The LCE films were prepared by in situ photopolymerization of an azobenzene LC monomer and a diacrylate crossliner containing an azobenzene moiety ($m = 5$, $n = 6$) in a glass cell at an LC temperature. Both of the glass slides of the cell were pre-coated with rubbed polyimide films as alignment layers. All the mesogens in LCEs were homogeneously aligned along the rubbing direction, as indicated by measurements of polarizing optical microscope (POM) and atomic force microscope (AFM).

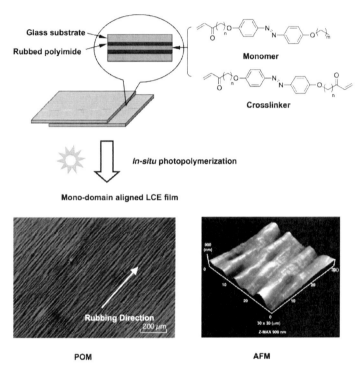

Figure 6.11 Preparation and morphologies of freestanding LCE films with monodomain-aligned mesogens using an azobenzene LC monomer and a crosslinker.

The resulted freestanding films of LCEs showed reversibly bending and unbending behaviors when they were by irradiated with unpolarized UV and then visible light, respectively. It was observed that the monodomain aligned LCE films bent toward the irradiation direction of the incident UV light along the rubbing direction, and the bent film reverted to the initial flat state after exposure to visible light, as shown in Fig. 6.12.

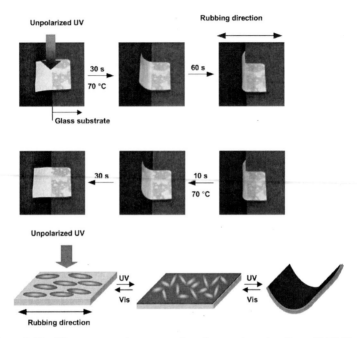

Figure 6.12 Photoresponsive properties of monodomain-aligned LCE films and their plausible mechanism of photoinduced bending. Reprinted with permission from Yu et al., Copyright 2004, American Chemical Society.

Since the molar extinction coefficient (ε) of azobenzene moieties at around 360 nm is large, photons are absorbed only in the surface of LCE films (a thickness of several tens of micrometers) and *trans–cis* photoisomerization occurs only in the surface area of these freestanding LCE films, which leads to an analogous bilayer structure with mesogenic slabs of two different polymers that respond differently, and "bending" can be explained as movement like a bimetal way [Yu et al., 2003]. As a result, the volume contraction is generated only in the surface layer, thus causing the LCE films to bend toward the light source. The plausible schematic illustration of the photoinduced bending is shown in Fig. 6.12. Since only a part of the polymer films is involved in deformation in these materials, one can induce bending much faster than other modes of photoinduced deformations. It has been demonstrated that both the alignment ordering of azobenzene mesogens and crosslinking density strongly influence the bending performances of the azobenzene-containing LCE films.

Moreover, the monodomain aligned LCE films show anisotropic photoinduced bending behaviors, as shown in Fig. 6.13. Bending and unbending behaviors were observed in an anisotropic way, which occurred only along the rubbing direction of the alignment layers since the photoinduced volume change should be obtained only in the alignment direction of the mesogens in LCE films.

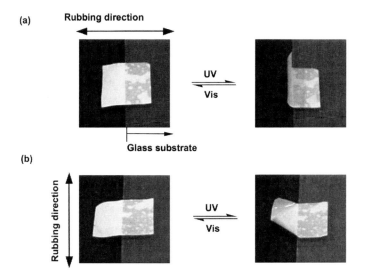

Figure 6.13 Anisotropic bending behavior of monodomain-aligned LCE films. Reprinted with permission from Yu et al., Copyright 2004, American Chemical Society.

6.3.2 Precise Control of Photoinduced Bending

As described in Chapter 3, anisotropic photoisomerization of azobenzene moieties in LCPs can be induced by means of the selective absorption of linearly polarized light (LPL), which can be used to precisely photomanipulate direction-selective movements of the LCE films. A variety of LCE films with different alignments of azobenzene mesogens were prepared and examined to elucidate the influence of the alignment on the photoinduced bending behavior. Yu et al. [2003] succeeded in achieving a photoinduced direction-controllable bending in single polydomain LCE films.

As shown in Fig. 6.14, only by changing the polarization direction of the actinic light, the bending of polydomain LCE films can be induced repeatedly and precisely along any chosen direction, and they bent toward the irradiation direction of the incident light with bending occurring parallel to the direction of LPL.

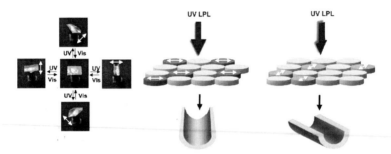

Figure 6.14 Photographs of precise control of the bending direction of polydomain LCE films by linearly polarized light (left) and the plausible mechanism (right).

Differently from the homogeneously aligned LCE films, homeotropically aligned ones were found to be bent away from the actinic light source upon UV exposure [Kondo et al., 2006]. As shown in Fig. 6.15, the alignment direction of azobenzene mesogens in the homeotropic-alignment LCE films is perpendicular to the film surface, and exposure to UV light induces an isotropic expansion, contributing to the bending behavior in a completely opposite way.

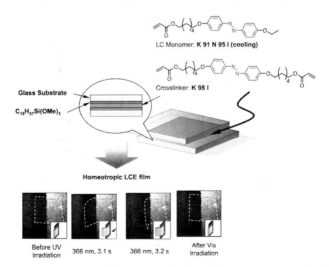

Figure 6.15 Photoinduced bending behavior of hometropically aligned LCE films. Reprinted with permission from Kondo et al., Copyright 2006, WILEY-VCH Verlag GmbH & Co. KGaA, Weinheim.

Upon UV irradiation, photomodulation of LCs occurs only in the surface region of LCE films because of the large molar extinction coefficient of photochromic LCs and the actinic light cannot penetrate through a thick film (thickness >10 μm). This induces the volume change only in film surface and generates photomechanical effect, as shown in Fig. 6.16. When the alignment of LC molecules is parallel to the surface of substrates, volume contraction is produced just along this pre-aligned direction, contributing to the anisotropic bending behavior toward a light source. On the contrary, volume expansion is brought about when the LC molecules are aligned perpendicularly to the substrates, resulted in different bending behavior, away from the light source. Furthermore, the photomechanical behavior is reversible if azobenzene molecules are used as photoresponsive mesogens.

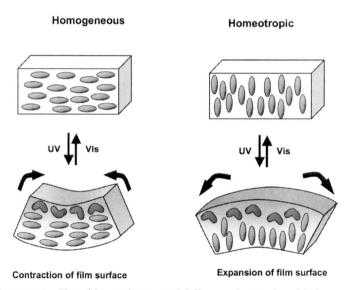

Figure 6.16 Plausible mechanism of different photoinduced behaviors of LCEs with homogeneous and hometropic alignment mesogens. Reprinted with permission from Kondo et al., Copyright 2006, WILEY-VCH Verlag GmbH & Co. KGaA, Weinheim.

In addition, LCE films with hybrid-aligned mesogens were prepared, in which a homogenous alignment was on one surface and a homeotropic alignment was on the other surface [Kondo et al., 2007]. Upon irradiation with unpolarized UV light from the side of the homogeneous surface, the hybrid LCE films bent toward the light source along the alignment direction, whereas the film bent away from the light source when the homeotropic surface was irradiated (Fig. 6.17). Furthermore, upon irradiation from both surfaces of the films, the bending speed was greatly enhanced at the same time comparing with homogeneously or homeotropically aligned LCEs. Recently, Harris et al. [2005] prepared LCE films with a twisted configuration of azobenzene moieties, showing a large amplitude bending motion and a large amplitude coiling motion upon exposure to UV light, which arises from the 90° twisted LC alignment configuration.

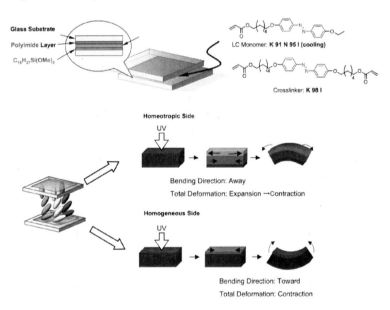

Figure 6.17 Photoresponsive behaviors of LCE films with hybrid-aligned mesogens.

6.3.3 Effect of Order of Mesogen on Photoinduced Bending

The order parameter of molecular orientation of mesogens is given in Fig 6.18, and the ferroelectric LCs are almost the highest in their order of mesogens. Besides, their molecular alignment can be controlled quickly by applying an electric field due to the presence of spontaneous polarization. Therefore, ferroelectric LCE films with a high LC order and a low glass transition temperature (T_g) were prepared by in situ photopolymerization of oriented LC mixtures of monomers and crosslinkers shown in Fig. 6.18, which were pre-aligned under an electric field [Yu et al., 2007]. Upon UV irradiation at room temperature, the films bent toward the actinic light source with a tilt angle to the rubbing direction of the polyimide alignment layer. This bending behavior was completed with 0.5 s of irradiation from a laser beam, demonstrating its quick photoresponse. The mechanical force generated by photoirradiation reached about 220 kPa, similar to the contraction force of human muscles (about 300 kPa).

Here, it is worth mentioning that these results expand our visions in photomechanical and photomobile effect of LCEs. However, several problems still remain unsolved for practical applications: the response time, the photodriven mechanism and the type of photoinduced deformation, which need further work in the aspects of engineering and materials. Recently, several novel materials for LCEs to improve their performance were explored.

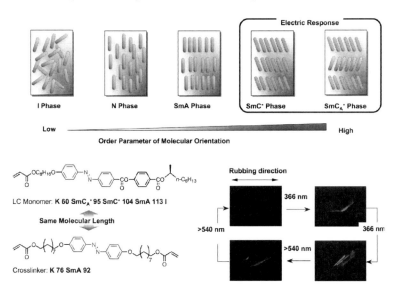

Figure 6.18 Order parameter of molecular orientation of mesogens (above) and photoresponsive behaviors of the ferroelectric LCE films (below). Reprinted with permission from Yu et al., Copyright 2007, WILEY-VCH Verlag GmbH & Co. KGaA, Weinheim.

6.3.4 Effect of Light Polarization on Photoinduced Bending

Tabiryan et al. [2005] prepared azobenzene-containing LCE films with homogeneously aligned mesogens and studied their photomechanical behaviors. They demonstrated that the LCE films exhibited an extraordinarily strong and fast mechanical response to a laser beam with a low intensity at room temperature. More interestingly, the photoinduced deformation showed a strong dependence of polarization of the actinic laser beam although the LCEs are composed of homogeneously aligned mesogens. The direction of the photoinduced bending or twisting of LCEs can be reversibly controlled by changing the polarization direction of the laser, which is attributed to the results of photoinduced reorientation of azobenzene moieties in the LCEs (Fig. 6.19).

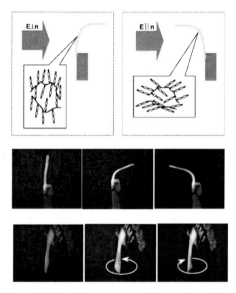

Figure 6.19 Effect of light polarization on photoinduced bending of LCEs.

6.3.5 Effect of LCE Structures on Photoinduced Bending

Undoubtedly, the LCE structure plays an important role in their photoresponsive behaviors. Upon preparation of LCEs, photoisomerizable azobenzene can either be included or not in the molecular structure of the crosslinkers. However, the resulted LCE films showed different bending behaviors for homogeneously aligned LCE films [Priimagi et al., 2012]. As shown in Fig. 6.20, when the azobenzenes are included in the crosslinkers, the film contracts upon UV irradiation and bends toward the light source. Under similar irradiation conditions, photoinduced expansion and bending away from the excitation light source is observed if an LCE film containing non-photoinert crosslinkers but the same concentration of the azobenzenes as side chains. This might originate from the competition between long-range uniaxial strain generated by the photoisomerization of the crosslinked azobenzene moieties (leading to photoinduced contraction) and the higher free-volume requirement of the *cis*-azobenzenes as compared to *trans*-azobenzenes (leading to photoinduced expansion).

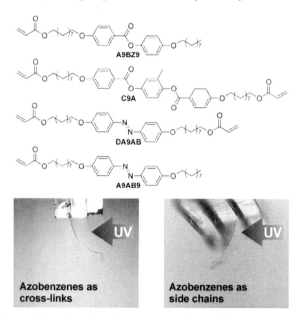

Figure 6.20 Effect of molecular structure of crosslinkers on photoinduced bending of homogeneously aligned LCEs.

In addition of the molecular structure of the used crosslinkers, the crosslinking degree of LCEs also influences their photoresponsive behaviors. As shown in Fig. 6.21, LCEs prepared with LC mixtures of a monomer and a crosslinker containing a spacer of undecylene showed different bending direction upon UV irradiation [Zhang et al., 2010]. When the temperature was lower than 90°C, the LCE film with a low crosslinking density first bent away from the light source and then toward it upon the irradiation of UV light. As the temperature was raised above 90°C, the film bent directly toward the light source. This interesting photoresponse was ascribed to the phase structure of the LCEs at different temperatures. The LCEs with a high crosslinking density bent directly toward the light source, but with a faster speed.

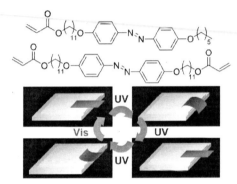

Figure 6.21 Photoresponse of LCEs with an LC monomer and crosslinker containing a spacer of undecylene. The crosslinking density influences the bending direction.

6.4 Microscale LCE Actuator

Combine with microscale techniques, miniature LCEs has found their applications as microactuators. Till now, several methods have been applied, such as microfluid [Ohm, Serra, & Zentel, 2009], microtemplates [Ohm et al., 2011], soft lithography [Buguin et al., 2006], and so on. The typical application of mciro-LCEs as microvalve for microfluidics was also reported [Sánchez-Ferrer & Fischl, 2011]. Here, we only concentrate on micro-LCEs showing photocontrollable properties.

6.4.1 LCE Fibers

It is well known that human's muscles are made of many bundles of muscle fibers and their anisotropic contractions are induced by electric stimuli. To construct artificial muscles, LCE fibers were fabricated due to their highly mechanical flexibility [Yoshino et al., 2010]. As shown in Fig. 6.22, LCE fibers with a high orientational order of mesogens along the LCE fiber axis were fabricated.

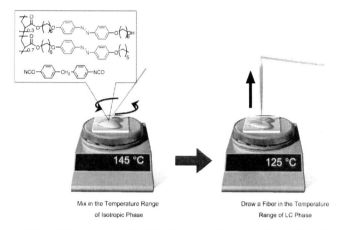

Figure 6.22 Fabrication of LCE fibers using urethane bonding for crosslinking.

Such LCE fibers showed good photoresponse upon exposure to UV light. Similarly to that of LCE films shown in Fig. 6.12, they bent toward the irradiation direction and reverted to the initial state upon exposure to visible light. One of the additional advantages of the LCE in fiber state is that flexibility of photoinduced 3D motions is greatly enhanced since the LCE fibers can be photodriven toward any desired direction (Fig. 6.23).

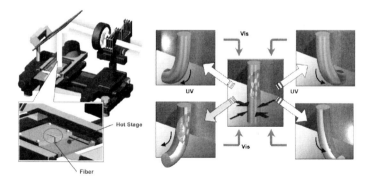

Figure 6.23 Photoresponse of the LCE fibers in 3D ways. Reprinted with permission from Yoshino et al., Copyright 2010, WILEY-VCH Verlag GmbH & Co. KGaA, Weinheim.

6.4.2 LCEs for Artificial Cilia

Recently, Oosten et al. [2009] combined LCEs with inkjet-printing approach and fabricated photorespoinsive microactuators. They utilized LC monomers as inks, which were inkjet-printed onto rubbed polyimide alignment layer deposited on a sacrificial PVA layer, as shown in Fig. 6.24. The LC self-assembling properties enables one to create large strain gradients, thus light-driven actuation is chosen to allow simple and remote addressing. By using multiple inks, microactuators with different subunits are created, which can be selectively addressed by changing the wavelength of the light. The actuators mimic the motion of natural cilia upon light irradiation.

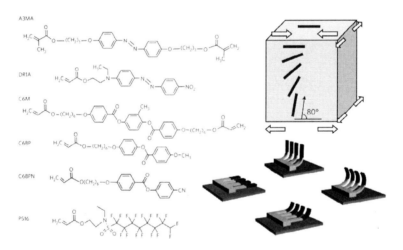

Figure 6.24   LCE microactuators mimic the motion of natural cilia.

6.4.3 LCEs with Micromolding

As shown in Fig. 6.25, a facile method for trapping the photoresponsive LCEs in surface microstructures via micropatterning was reported [Yang et al., 2006]. The confinement of LCE materials into surface monodomains enables the formation of reversible, shape-shifting surface patterns. Upon irradiation of UV light, the LCE monodomains were photocaused into an isotropic state. This causes the features switch between imprinted circular features and anisotropic liquid-crystalline features (AFM image in Fig. 6.25).

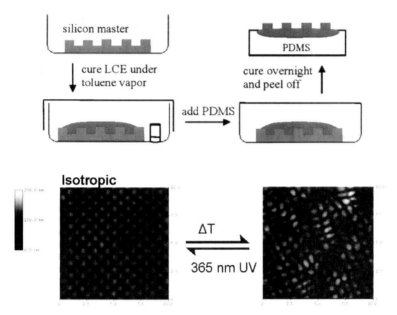

Figure 6.25 Photoresponsive LCE microactuators prepared with micropatterning. Reprinted with permission from Yang et al., Copyright 2006, American Chemical Society.

6.4.4 Photocontrol of Surface-Relief Formation

In Chapter 5, surface-relief structures based on light-driven materials transport have been introduced upon irradiation with interference light from two laser beams. For LCEs, surface relief also can be modulated at a low light intensity. The principle is based on anisotropic geometric changes of LCEs upon a change of the molecular order parameters, as shown in Fig. 6.26 [Li et al., 2012]. The pattern was obtained by alternating the homeotropic alignment and the planar cholesteric LC orientation. Upon UV irradiation, expansion occurs in the vertical alignment area. In the planar cholesteric LC area, where the molecules are oriented on average planar to the surface, the reduction of the order parameter results in a positive expansion normal to the plane, whereas the expansion in the plane is close to zero or even negative. This leads to almost no volume changes in the patterning direction. Thus, pressure gradient was produced between the two areas, leading to dynamically photocontrol the surface topology.

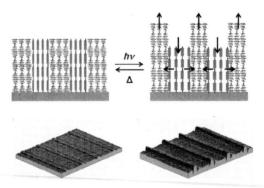

Figure 6.26 Photocontrol surface relief by light-induced molecular order parameters. Reprinted with permission from Liu et al., Copyright 2012, WILEY-VCH Verlag GmbH & Co. KGaA, Weinheim.

6.5 Novel LCE Materials and Photomechanical Ways

6.5.1 Recyclable Hydrogen-Bonded LCEs

Similar to chemically crosslinked LCE films, the photomechanical effect was also obtained in freestanding LCE films composed of hydrogen-bonded supramolecularly crosslinking of an azobenzene-containing copolymer and an LMW crosslinker [Mamiya et al., 2008]. As shown in Fig. 6.27, the LMW crosslinker acts as both roles of photoresponsive function and hydrogen-bond acceptor due to the existence of pyridyl groups at both ends.

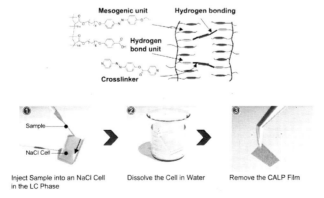

Figure 6.27 Preparation of the hydrogen-bonded LCE film using NaCl cell.

The photoinduced bending and unbending behaviors of the hydrogen-bonded LCE film are similar to that of the covalently bonded LCE films. This means that the crosslinks formed by hydrogen bonds can convert the motion of the mesogens into a macroscopic change of the LCE films, which is similar to the covalently bonded ones. This kind of supramolecularly self-assembled LCE films could be reconstructed through a recycle process shown in Fig. 6.28, which is superior to the covalently bonded materials.

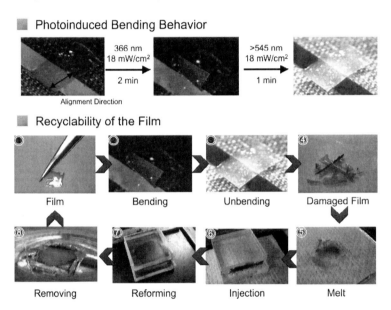

Figure 6.28 Photoinduced bending behavior and their recyclability of the hydrogen-bonded LCE film.

6.5.2 Dye-Doped LCEs

Since the first report of photoinduced 3D motion of LCEs, several other interesting movements of photoresponsive LCEs were also reported. Camacho-Lopez et al. [2004] demonstrated that mechanical deformation of an LCE sample doped with azobenzene compounds becomes very large in response to non-uniform visible-light illumination. As shown in Fig. 6.29, when a laser beam is irradiated from above onto a dye-doped LCE sample floating on water, the LCE swims away from the laser beam, with an action resembling that of flatfish. They obtained more than 60° bending in the dye-doped LCEs.

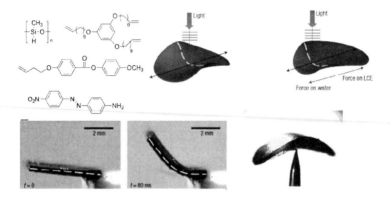

Figure 6.29 Photoresponse of azobenzene-dye doped LCEs.

6.5.3 Twisted LCEs

Harris et al. [2005] prepared LCE films with a densely crosslinked, twisted configuration of azobenzene moieties. The films showed a large amplitude bending and 55 coiling motion upon exposure to UV light which results from the twisted configuration of the LC alignment (Fig. 6.30).

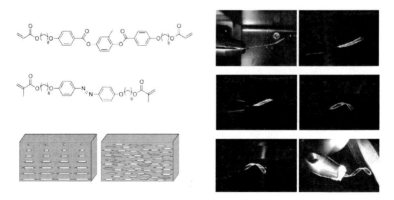

Figure 6.30 Photoresponse of twisted and uniaxial LCE films (left) and UV-induced coiling of a film in the twisted configuration (right).

6.5.4 Hummingbird Movement

The photoisomerization of azobenzenes can be induced in a high speed (picosecond). When they are confined in LCEs by crosslinked networks, high frequency and large amplitude oscillations have been observed, as shown in Fig. 6.31 [White et al., 2008]. Driven by laser exposure, the photosensitive LCEs in a cantilever shape show fast (30 Hz) and large amplitude (>170°) oscillation. The frequency of the photodriven oscillation is similar to a hummingbird wing beat, which can range from 20–80 Hz. Such photoinduced oscillation can be turned on and off by switching the polarization direction of the actinic laser beam, and the behavior shows little fatigue over 2,50,000 cycles.

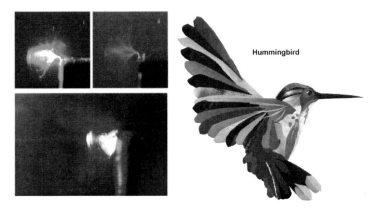

Figure 6.31 Photoinduced oscillation of LCEs in a high frequency similar to a hummingbird wing beat.

Besides, much effort has been paid to the actinic light with a longer wavelength. Both visible light and near infrared light (NIR) have been successfully applied in photodriving the movements of LCEs [Cheng et al., 2010a, 2010b; Wu et al., 2011].

6.6 LCE-Laminated Films

Generally, the 3D photomanipulation of LCE films can be greatly accelerated by increasing the experimental temperature. Below T_gs of LCE materials, the photomechanical movements are usually very slow and sometimes difficult to be observed. By molecular design, photoresponsive behavior at room temperature was obtained in LCE films prepared with an azobenzene-containing monomer and a crosslinker (m = 8, n = 9 in Fig. 6.11). Benefit from this, diverse complicated 3D movements like an inchworm walk and a flexible robotic arm motion as well as photodriven plastic motor were explored very recently, which is a completely photomobile performance of LCEs [Yamada et al., 2008, 2009].

Comparing with photoinduced 3D motions of bending and unbending behavior, photodriven rotation is undoubtedly more challengeable from an engineering point of view. Yamada et al. [2008] first developed light-induced rotation based on azobenzene-containing LCE films at room temperature. They prepared continuous rings by connecting both ends of the LCE films, in which azobenzene mesogens were homogeneously aligned along the circular direction of the rings (Fig. 6.32). Upon simultaneous irradiation with UV light (from the downside right) and visible light (from the upside right), the ring rolled intermittently toward the actinic light source, resulting almost in a 360° roll at room temperature.

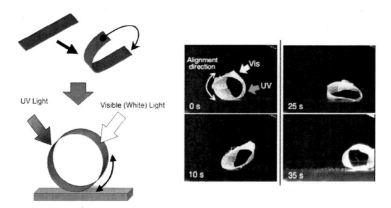

Figure 6.32 Preparation of continuous ring of LCE films and their rolling motion.

Figure 6.33 gives the possible mechanism of the photodriven rotation of the LCE film rings. Upon exposure to UV light, a local contraction force is generated in the irradiated areas of the belt along the alignment direction of the azobenzene mesogens, which is parallel to the long axis of the belt. At the same time, by irradiation with visible light, a local expansion force is produced at the irradiated parts of the belt along the alignment direction in the belt (also along the long axis of the belt). These contraction and expansion forces produced simultaneously at the different parts along the long axis of the belt could induce the rotation of the belt. This rotation of the belt makes the pulleys rotate in the same direction concomitantly. The rotation brings new parts of the belt to be exposed to UV and visible light, which enables the belt and pulleys to rotate continuously.

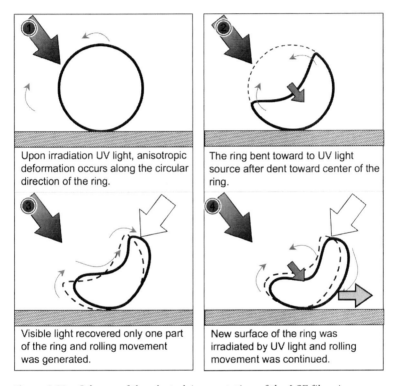

Figure 6.33 Scheme of the photodriven rotation of the LCE film rings.

To improve the mechanical properties of azobenzene-containing LCEs, they prepared a plastic belt of the LCE-laminated films by attaching LCE films on flexible polyethylene (PE) sheet, and then placed the belt on a homemade pulley system as illustrated in Fig. 6.34 [Yamada et al., 2009]. As a flexible plastic film, a low-density PE unstretched film was used because of its good flexibility and mechanical properties at room temperature. The LCE laminated films were prepared by thermal compression bonding of an LCE layer and a PE film with an adhesion layer.

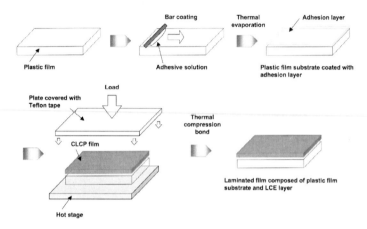

Figure 6.34 Fabrication of LCE-laminated films.

Figure 6.35 shows the photoresponsive behaviors of the LCE laminated film prepared in Fig. 6.34 [Yamada et al., 2009]. In this LCE laminated film, the azobenzene mesogens were aligned along the long axis of the film. The film extended to a flat shape by irradiation with UV light and reverted to the initial bent shape by irradiation with visible light (>540 nm), which could be repeated at room temperature just by changing the wavelength of the actinic light. Upon exposure to UV light, the *trans–cis* photoisomerization of the azobenzene moieties occurs only on the surface of the LCE layer and a contraction force is generated along the alignment direction of the azobenzene mesogens, which is parallel to the long axis of the LCE laminated film, and the contraction of the LCE layer results in extension of the whole LCE laminated film. The *cis–trans* back-isomerization occurs by irradiation with visible light and the film reverted to the initial bent state. Here, it is worth mentioning that an enough mechanical force can be generated in the LCE layer partly laminated, only about 25% as large as the PE film, to move the whole film drastically by photoirradiation. In other words, we can transform ordinary plastic films into photomobile materials just by laminating LCE layers partly according to need. Furthermore, the LCE laminated films have good mechanical strength properties and can be easily fabricated into a diversified form and an arbitrary size due to good processing properties of plastic films.

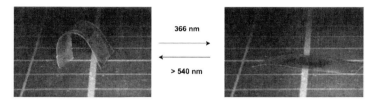

Figure 6.35 Photoresponsive behaviors of the LCE laminated film.

By carefully designing the LCE laminated film, more complicated motions have been obtained. For instance, Fig. 6.36 demonstrates a unidirectional motion, an inchworm walk, of the LCE laminated film with asymmetric end shapes. The film moved forward by alternate irradiation with UV and visible light at room temperature. A plausible mechanism of this motion is illustrated in the right of Fig. 6.36 [Yamada et al., 2009]. The film has an asymmetrical shape: one end is a sharp edge and the other is flat. Upon exposure to UV light, the film extends forward because the sharp edge acts as a stationary point, and the film retracts from the rear side by irradiation with visible light because the flat edge acts as a stationary point, which enables the film to move in one direction. Besides, the LCE-laminated films exhibited other photomobility of sophisticated 3D motions like inchworm walk and a flexible robotic arm.

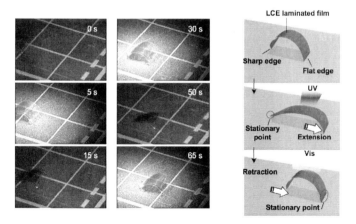

Figure 6.36 Photodriven movement of the LCE laminated film.

A motor actuator is one of the most useful energy conversion systems. In addition to the 3D motions, a continuous plastic belt of the LCE laminated film by connecting both ends of the film and put the belt on a homemade pulley system. As shown in Fig. 6.37, upon irradiation of the belt with UV light from top right and visible light from the top left simultaneously, a rotation of the belt was induced to drive the two pulleys in a counterclockwise direction at room temperature. This is the first realization of light-driven plastic motors, which directly converts light to mechanically rotational energy with soft materials. It is believed that the sections that were exposed to light expand while those regions away from the light contract, generating an overall rotating moment of the plastic films (Fig. 6.37). A reverse rotation of this belt was also obtained just by changing the irradiation positions of the UV and visible light.

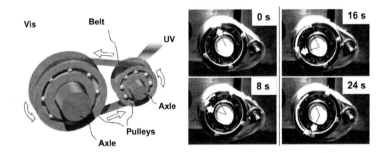

Figure 6.37 Schematic illustration of a light-driven plastic motor and photographs of time profiles of the rotation.

As light can be handled remotely, instantly and precisely, these plastic materials can work as main driving parts of the light-driven actuators without the aid of batteries, electric wires and gears. Recently, a novel method to prepare LCE laminated film by directly coating an azobenzene-containing LCP homopolymers on a flexible polymer substrate. The azobenzene-containing LCP was pre-aligned by a linearly polarized laser beam and the following crosslinked by electron beams [Naka et al., 2011]. Such EB-crosslinked azobenzene LCPs could successfully work as photomobile polymer materials with an adhesive-free bilayer structure. In addition, the durability in the adhesive-free bilayer film was much improved as compared with the laminated films. This simple method can be used for fabrication of photomobile polymer materials with large-area, smooth surface, and controlled thickness of both photoactive and substrate layers.

For an easy preparation of photocontrollable LCE elastomers, the synthesis of an azobenzene monomer and an azobenzene crosslinker shown in Fig. 6.11 was improved [Li et al., 2009]. Recently, several photoinduced motion or deformation was found in novel systems containing azobenzene chromophores. More interestingly, reversible mechanical bending of plate-like microcrystals of a LMW azobenzene compound was induced upon UV irradiation [Koshima et al., 2009]. Besides, such reversible photomechanical effects of LCE films have been observed in the glassy state [Lee et al., 2011]. Such novel materials with light-induced motion and deformation would be helpful to elucidate the photocontrollable performances of LCE elastomers.

References

Broer, D. J, Boven, J., Mol, G. N., and Challa, G. (1989). In-situ photopolymerization of oriented liquid-crystalline acrylates, *Makromol. Chem.* **190**, pp. 2255–2268.

Buguin, A., Li, M. H., Silberzan, P., Ladoux, B., and Keller, P. (2006) Micro-actuators: When artificial muscles made of nematic liquid crystal elastomers meet soft lithography, *J. Am. Chem. Soc.* **128**, pp. 1088–1089.

Camacho-Lopez, M., Finkelmann, H., Palffy-Muhoray, P., and Shelley, M. (2004). Fast liquid-crystal elastomer swims into the dark, *Nat. Mater.* **3**, pp. 307–310.

Cheng, F., Yin, R., Zhang, Y., Yen, C., and Yu, Y. (2010a). Full plastic microrobots: Manipulate objects only by visible Light, *Soft Matter* **6**, pp. 3447–3449.

Cheng, F., Zhang, Y., Yin, R., and Yu, Y. (2010b). Visible light induced bending and unbending behavior of crosslinked liquid-crystalline polymer films containing azotolane moieties, *J. Mater. Chem.* **20**, pp. 4888–4896.

De Gennes, P. G. (1975). Physique moleculaire, *C. R. Acad. Sci. B* **281**, pp. 101–103.

Eisenbach, C. D. (1980). Isomerization of aromatic azo chromophores in poly(ethyl acrylate) networks and photomechanical effect, *Polymer* **21**, pp. 1175–1179.

Finkelmann, H., Kock, H. J., and Rehage, H. (1981). Investigations on liquid crystalline polysiloxanes.3. Liquid crystalline elastomers—a new type of liquid crystalline material, *Makromol. Chem. Rapid Commun.* **2**, pp. 317–322.

Finkelmann, H., Nishikawa, E., Pereira, G. G., and Warner, M. (2001). A new opto-mechanical effect in solids, *Phys. Rev. Lett.* **87**, pp. 15501(1–4).

Harris, K. D., Cuypers, R., Scheibe, P., van Oosten, C. L., Bastiaansen, C. W. M., Lub, J., and Broer, D. J. (2005). Large amplitude light-induced motion in high elastic modulus polymer actuators, *J. Mater. Chem.* **15**, pp. 5043–5048.

Hogan, P. M., Tajbakhsh, A. R., and Terentjev, E. M. (2002). UV manipulation of order and macroscopic shape in nematic elastomers, *Phys. Rev. E* **65**, pp. 041720 (1–10).

Ikeda, T., Nakano, M., Yu, Y., Tsutsumi, O., and Kanazawa, A. (2003). Anisotropic bending and unbending behavior of azobenzene liquid-crystalline gels by light, *Adv. Mater.* **15**, pp. 201–205.

Ikeda, T., Mamiya, J., and Yu, Y. (2007). Photomechanics of liquid-crystalline elastomers and other polymers, *Angew. Chem. Int. Ed.* **46**, pp. 506–528.

Kondo, M., Yu, Y., and Ikeda, T. (2006). How does the initial alignment of mesogens affect the photoinduced bending behavior of liquid-crystalline elastomers, *Angew. Chem. Int. Ed.* **45**, pp. 1378–1382.

Kondo, M., Yu, Y., Mamiya, J., Kinoshita, M., and Ikeda, T. (2007). Photoinduced deformation behavior of crosslinked azobenzene liquid-crystalline polymer films with unimorph and bimorph structure, *Mol. Cryst. Liq. Cryst.* **478**, pp. 245–257.

Koshima, H., Ojima, N, and Uchimoto, H. (2009). Mechanical motion of azobenzene crystals upon photoirradiation, *J. Am. Chem. Soc.* **131**, pp. 6890–6891.

Kurihara, S., Sakamoto, A., and Nonaka, T. (1998). Fast photochemical switching of a liquid-crystalline polymer network containing azobenzene molecules, *Macromolecules* **31**, pp. 4648–4650.

Kurihara, S., Sakamoto, A., Yoneyama, D., and Nonaka, T. (1999). Photochemical switching behavior of liquid crystalline polymer networks containing azobenzene molecules, *Macromolecules* **32**, pp. 6493–6498.

Lee, K., Koerner, H., Vaia, R. A., Bunninga, T. J., and White, T. J. (2011). Light-activated shape memory of glassy, azobenzene liquid crystalline polymer networks, *Soft Matter* **7**, pp. 4318–4324.

Li, C., Lo, C., Zhu, D., Li, C., Liu, Y., and Jiang, H. (2009). Synthesis of a photoresponsive liquid-crystalline polymer containing azobenzene, *Macromol. Rapid Commun.* **30**, pp. 1928–1935.

Li, M. H., Keller, P., Li, B., Wang, X., and Brunet, M. (2003). Light-driven side-on nematic elastomer actuators, *Adv. Mater.* **15**, pp. 569–572.

Li, M. H., Keller, P., Yang, J., and Albouy, P. (2004). An artificial muscle with lamellar structure based on a nematic triblock copolymer, *Adv. Mater.* **16**, pp. 1922–1925.

Li, D., Bastiaansen, C. W. M., den Toonder, J. M. J., and Broer, D. J. (2012). Photo-switchable surface topologies in chiral nematic coatings, *Angew. Chem. Int. Ed.* **51**, pp. 892–896.

Lovrien, R. (1967). The photoviscosity effect, *Proc. Natl. Acad. Sci. USA* **57**, pp. 236–242.

Mamiya, J., Yoshitake, A., Kondo, M., Yu, Y., and Ikeda, T. (2008). Is chemical crosslinking necessary for the photoinduced bending of polymer films, *J. Mater. Chem.* **18**, pp. 63–65.

Naka, Y., Mamiya, J., Shishido, A., Washio, M., and Ikeda, Y. (2011). Direct fabrication of photomobile polymer materials with an adhesive-free bilayer structure by electron-beam irradiation, *J. Mater. Chem.* **21**, pp. 1681–1683.

Ohm, C., Serra, C., and Zentel, R. (2009). A continuous flow synthesis of micrometer-sized actuators from liquid crystalline elastomers, *Adv. Mater.* **21**, pp. 4859–4862.

Ohm, C., Haberkorn, N., Theato, P., and Zentel, R. (2011). Template-based fabrication of nanometer-scaled actuators from liquid-crystalline elastomers, *Small* **7**, pp. 194–198.

Oosten, C. L., Bastiaansen, C. W. M., and Broer, D. J. (2009). Printed artificial cilia from liquid-crystal network actuators modularly driven by light, *Nat. Mater.* **8**, pp. 677–682.

Priimagi, A., Shimamura, A., Kondo, M., Hiraoka, T., Kubo, S., Mamiya, J., Kinoshita, M., Ikeda, T., and Shishido, A. (2012). Location of the azobenzene moieties within the cross-linked liquid-crystalline polymers can dictate the direction of photoinduced bending, *ACS Macro Lett.* **1**, pp. 96–99.

Sánchez-Ferrer, A., Fischl, T., Stubenrauch, M., Albrecht, A., Wurmus, H., Hoffmann, M., and Finkelmann, H. (2011). Liquid-crystalline elastomer microvalve for microfluidics, *Adv. Mater.* **23**, pp. 4526–4530.

Tabiryan, N., Serak, S., Dai, X. D., and Bunning, T. (2005). Polymer film with optically controlled form and actuation, *Opt. Express* **13**, pp. 7442–7448.

White, T. J., Tabiryan, N. V., Serak, S. V., Hrozhyk, U. A., Tondiglia, V. P., Koerner, H., Vaia, R. A., and Bunning, T. J. (2008). A high frequency photodriven polymer oscillator, *Soft Matter* **4**, pp. 1796–1798.

Wu, W., Yao, L., Yang, T., Yin, R., Li, F., and Yu, Y. (2011). NIR-light-induced deformation of cross-linked liquid-crystal polymers using upconversion nanophosphors, *J. Am. Chem. Soc.* **133**, pp. 15810–15813.

Yamada, M., Kondo, M., Mamiya, J., Yu, Y., Kinoshita, M., Barrett, C. J., and Ikeda, T. (2008). Photomobile polymer materials—towards light-driven plastic motors, *Angew. Chem. Int. Ed.* **47**, pp. 4986–4988.

Yamada, M., Kondo, M., Miyasato, R., Naka, Y., Mamiya, J., Kinoshita, M., Shishido, A., Yu, Y., Barrett, C. J., and Ikeda, T. (2009). Photomobile polymer materials - various three-dimensional movements, *J. Mater. Chem.* **19**, pp. 60–62.

Yoshino, T., Kondo, M., Mamiya, J., Kinoshita, M., Yu, Y., and Ikeda, T. (2010). Three-dimensional photomobility of crosslinked azobenzene liquid-crystalline polymer fibers, *Adv. Mater.* **22**, pp. 1361–1363.

Yu, Y., Nakano, M., and Ikeda, T. (2003). Directed bending of a polymer film by light, *Nature* **425**, pp. 145–145.

Yu, Y., Maeda, T., Mamiya, J., and Ikeda, T. (2007). Photomechanical effects of ferroelectric liquid-crystalline elastomers containing azobenzene chromophores, *Angew. Chem. Int. Ed.* **46**, pp. 881–883.

Yang, Z. Q., Herd, G. A., Clarke, S. M., Tajbakhsh, A. R., Terentjev, E. M., and Huck, W. T. S. (2006). Thermal and UV shape shifting of surface topography, *J. Am. Chem. Soc.* **128**, pp. 1074–1075.

Zhang, Y., Xu, J., Cheng, F., Yin, R., Yen, C., and Yu, Y. (2010). Photoinduced bending behavior of crosslinked liquid-crystalline polymer films with a long spacer, *J. Mater. Chem.* **20**, pp. 7123–7130.

Chapter 7

Liquid-Crystalline Block Copolymers

Recent progress in materials chemistry allows us to freely design advanced materials with almost arbitrarily controllable multifunctions. Liquid-crystalline block copolymers (LCBCs) are one of such kinds of materials, which elegantly integrate LCs and block copolymers (BCs), the two kinds of naturally self-assembled soft materials into one organic system [Gallot, 1996; Walther & Finkelmann, 1996]. The combination of LCs and BCs provides LCBCs with novel functionalities as well as the possibility of nanostructure formation and control at the existence of the LC ordering. Being among novel types of macromolecules of industrial and academic interest, the fascinating LCBCs have been extensively studied for their potential applications in biology, photonics, nanotemplates, and nanofabrication processes.

Due to their immiscible properties, LCBCs bearing at least one mesogenic block often microphase separates into varieties of nanostructures like sphere, cylinder, and lamellae phase domains with an increase in volume ratio of the LC blocks (Fig. 7.1). LCBCs also inherit LC performances such as self-organization, long-range ordered fluidity, molecular cooperative motion, formed large birefringence, anisotropy in various physical properties (optical, electrical, and magnetic fields), and alignment change by external fields at surfaces and interfaces. These provide the designed LCBCs

---

Dancing with Light: Advances in Photofunctional Liquid-Crystalline Materials
Haifeng Yu
Copyright © 2015 Pan Stanford Publishing Pte. Ltd.
ISBN 978-981-4411-11-0 (Hardcover), 978-981-4411-12-7 (eBook)
www.panstanford.com

with unique features different from traditional amorphous BCs, and enable them to attract much attention in macromolecular engineering.

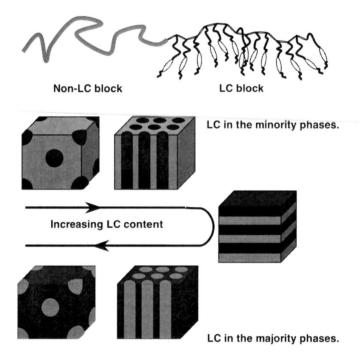

Figure 7.1 Schematic illustration of microphase separation in photoresponsive LCBCs.

7.1 Synthesis of Well-Defined LCBCs

The inherent microphase separation of LCBCs provides a convenient and economic method to fabricate regular nanostructures by top–down nanotechnology. To show regularly ordered nanostructures, LCBCs should have well-defined structures and low polydispersities of molecular weight. Moreover, each block of BCs has to be larger than a certain minimum molecular weight. Therefore, several polymerization methods, such as anionic, cationic, free radical, and metal-catalyzed polymerization have been explored to synthesize LCBCs that meet these requirements.

7.1.1 Direct Polymerization

Direct polymerization of chromophore-containing monomers via living processes is one of the most effective ways to synthesize well-defined LCBCs. In this method, a mono-dispersed macroinitiator should be firstly prepared, which is then used as a macroinitiator for the subsequent polymerization of chromophore-containing monomers, as shown in Fig. 7.2.

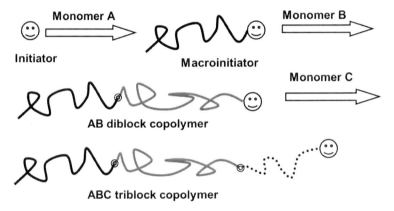

Figure 7.2   Preparation of well-defined LCBCs by direction polymerization.

In 1994, photoresponsive AB-type diblock copolymers were first synthesized by direct anionic polymerization of an azobenzene monomer [Finkelmann & Bohnert, 1994]. As shown in Fig 7.3, polymerization of polystyrene (PS)-based diblock copolymers was carried out from a PS-lithium capped with 1,1-diphenylethylene, while the poly(methyl methacrylate) (PMMA)-based diblock copolymers were prepared by addition of MMA monomers to the "living" azobenzene polyanion, obtained by reaction of 1,1-diphenyl-3-methylpentyl lithium with an azobenzene monomer in tetrahydrofuran (THF) at a lower temperature. By this method, a series of well-defined LCBCs were prepared with controlled molecular weights and narrow polydispersities [Lehmann et al., 2000].

In addition to the living anionic polymerization technique, radical processes have been used for synthesis of most of commercially available polymers and copolymers. The major success was the large number of monomers could undergo free radical polymerization in a convenient temperature range with the minimal requirement for purification of monomers and solvents. Undoubtedly, atom transfer radical polymerization (ATRP) method is one of the most popular among controlled or living radical polymerization. Its products often possess well-defined structures and narrow molecular-weight polydistributions [Matyjaszewski & Xia, 2001]. Since ATRP allows for a control over the chain topology, the composition and the end functionality for a large range of radically polymerizable monomers. Many LCBCs with specified structures have been synthesized by this approach [Yu et al., 2005a; 2005c; 2006a; 2006b; Tian et al., 2002].

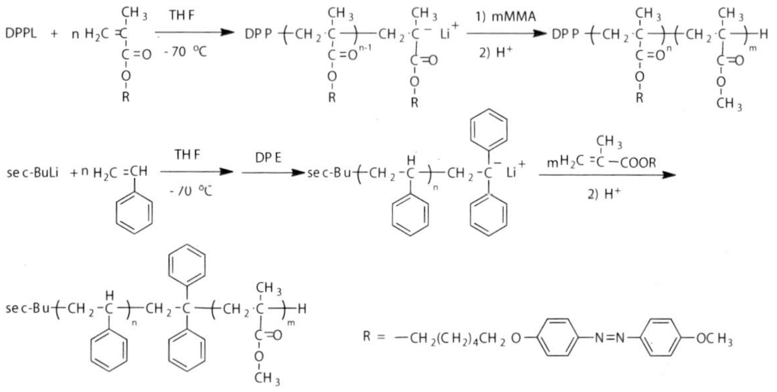

Figure 7.3 Direct preparation of LCBCs by anionic living polymerization. Reprinted with permission from Bohnert and Finkelmann, Copyright 1994, Hüthig & Wepf Verlag, Basel.

Recently, a modified ATRP method was developed to prepare novel amphiphilic LCBCs consisting of a flexible poly(ethylene oxide) (PEO) as hydrophilic segment and a poly(methacrylate) containing an azobenzene moiety in side chain as hydrophobic LC segment [Yu et al., 2006a; 2006b]. The unique characteristic of the obtained photoresponsive LCBCs was the nanoscaled microphase separation, in which a regular array of PEO nanocylinder (10–30 nm) with a periodicity of about 20–50 nm dispersed in a smectic LC polymer matrix. Starting from commercially available PEO, Yu et al. applied this approach to prepare several ABC-type LCBCs with photocontrol performances, as shown in Fig. 7.4 [Yu et al., 2005a; 2007a; 2007b; 2007e].

Synthesis of Well-Defined LCBCs | 215

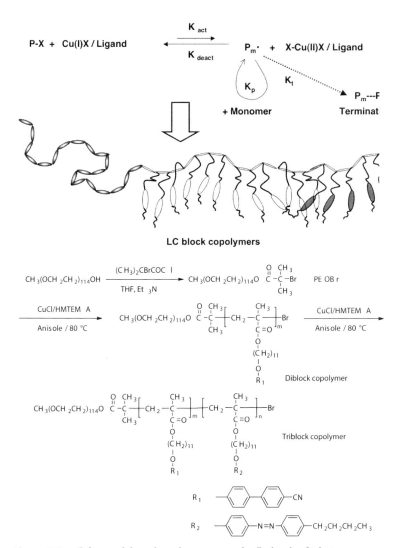

Figure 7.4 Scheme (above) and one example (below) of photoresponse LCBCs preparation of by a modified ATRP method. Reproduced with permission from John Wiley and Sons Publisher.

Besides ATRP, other controlled/living radical polymerization techniques such as reversible addition/fragmentation chain transfer polymerization (RAFT) [Zhao et al., 2008] and nitroxide-mediated polymerization (NMP) [Yoshida & Ohta, 2005] was also explored to synthesize LCBCs with designed molecular constitution.

7.1.2 Post-Functionalization

The LCBCs can also be prepared by post functionalization of active groups in block copolymer precursors, as shown in Fig. 7.5. In 1989, LCBCs with cholesteryl groups were first prepared by post-polymerization reaction [Adams & Gronski, 1989]. Then, a family of well-defined LCBCs was synthesized by a polymer analogue reaction starting from poly(styrene-b-isoprene) with a high content of pendent vinyl (and methyl vinyl) [Mao et al., 1997]. Quantitative hydroboration chemistry was used to convert the pendent double bonds of the isoprene block to hydroxyl groups, to which the photoresponsive groups were attached by acid chloride coupling. Due to the many possibilities to functionalize the hydroxyl group in poly(2-hydroxyethyl methacrylate), it was used to prepare LCBCs by other groups using a similar polymer analogue way [Frenz et al., 2004]. Recently, a post Sonogashira cross-coupling reaction of a reactive polymer precursor was used to prepare highly birefringent photoresponsive LCPs. This seemed to be one of candidate reactions to synthesized well-defined LCBCs [Yu et al., 2009c].

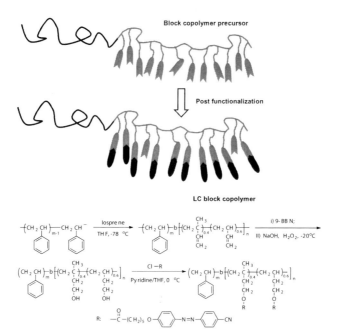

Figure 7.5 Schematic illustration (above) and one example (below) of LCBCs prepared by post-functionalization.

7.1.3 Supramolecular Self-Assembly

Upon supramolecular self-assembly, LMW additives can be used to adjust the properties of BCs by hydrogen bonding. This concept was first introduced in a PS-*b*-poly(4-vinylpyridine) (P4VP), which was stoichiometrically complex with pentadecylphenol molecules to form supramolecular complexes [Makinen et al., 2000]. Following this concept, well-ordered nanostructures were fabricated by supramolecular assembly of PS-*b*-P4VP and 2-(4-hydroxyazobenzene) benzoic acid (HABA), consisting of nanocylinders formed by P4VP–HABA associates in the PS matrix [Sidorenko et al., 2003] as shown in Fig. 7.6. Extraction of HABA with a selective solvent resulted in nanochannel membranes with a hexagonal lattice of hollow channels in the nanocylinders crossing the membrane from the top to the bottom. In addition, supramolecular self-assembly between PS-*b*-poly(acrylic acid) (PAA) and imidazobenzenele-terminated hydrogen-bonding mesogenic groups was also used to prepare LCBCs [Chao et al., 2004]. Owing to the attached LC properties, the nanostructures in the obtained LC block copolymer could be oriented by using an alternating-current (AC) electric field, in a direction parallel to the electrodes.

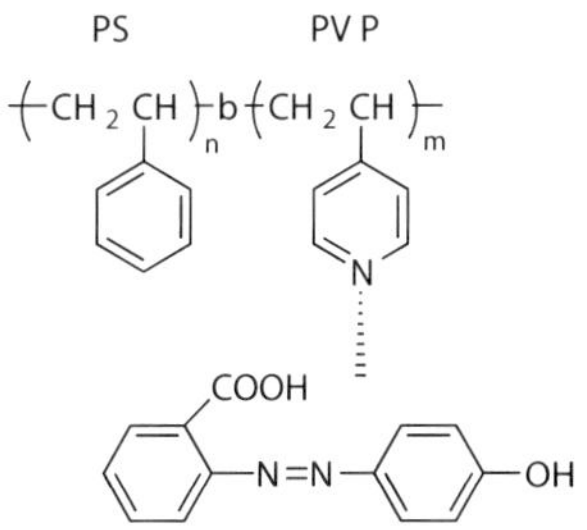

Figure 7.6 Preparation of photoresponsive BCs by supramolecular self-assembly between a BC and low-molecular-weight azobenzene compound. (Reproduced with permission from Sidorenko et al., 2003.)

7.1.4 Special Reactions

Some special reactions with a particularly designed route have been used to synthesize LCBCs, and such a reaction route includes at least two polymerization processes. As shown in Fig. 7.7, an initiator with a specified structure is used to prepare a macroinitiator with only one initiator in one polymer chain. Under certain conditions like light irradiation or thermal treatment, macroradicals can be formed by the decomposition of the macroinitiator. This induced additional radical polymerization of a photoresponsive mesogenic monomer from the decomposed points in the segment. Thus, AB-type or ABA-type photoresponsive LCBCs can be obtained by termination of the macroradicals (above scheme in Fig. 7.7).

For instance, a series of poly(vinyl ether)-based LCBCs were synthesized by using living cationic polymerization and free-radical polymerization techniques [Serhatli & Serhatli, 1998]. 4.4'-Azobenzenebis(4-cyano pentanol) (ACP) was used to quantitatively couple two well-defined polymers of living poly(vinyl ether), initiated by the methyl trifluoromethane sulfonate–tetrahydrothiophene system. Then, the ACP in the main chain was thermally decomposed to produce polymeric radicals, which was then used to initiate the polymerization of methyl methacrylate (MMA) or styrene to prepare PMMA-based or PS-based LCBCs (AB or ABA types). Although the obtained LCBCs showed narrow polydistributions, there is still no report about ABC-type LCBCs prepared by such a reaction route.

7.2 Phase Diagram of LCBCs

Figure 7.8 shows a universal phase diagram of symmetric AB-type BCs calculated using self-consistent mean-field theory [Matsen & Bates, 1996]. Here, χ is Flory–Huggins–Staverman interaction parameter, N is degree of polymerization, and f represents volume fractions of each block. Because of the additional LC properties, the phase diagram of LCBCs is a little different from Fig. 7.8 because several factors such as conformational asymmetry and structural asymmetry greatly influence their phase-segregated behaviors. Furthermore, the additional anchoring of the LC ordering to the intermaterial dividing surface (IMDS, Fig. 7.9) can affect the self-assembly behaviors of LCBCs [Osuji et al., 1999]. Therefore, the driving force for microphase separation of LCBCs should include the following items:

(1) Minimizing the interfacial energy
(2) Maximizing the conformational entropy of polymers

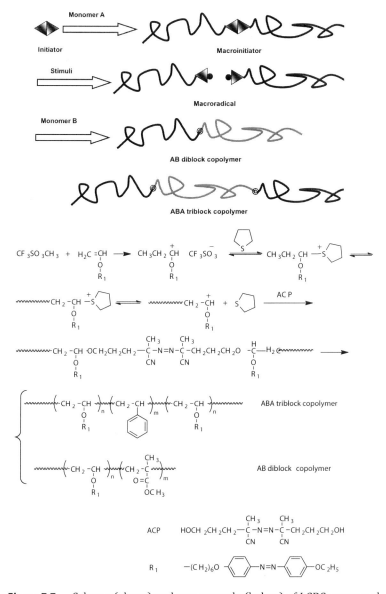

Figure 7.7 Scheme (above) and one example (below) of LCBCs prepared from a special reaction.

(3) Minimizing the additional elastic deformation of the LC phases.

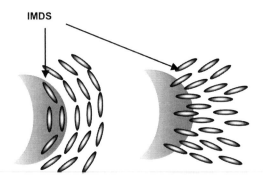

Homogeneous   homeotropic

Figure 7.8 Phase diagram of AB-type diblock copolymers calculated using self-consistent mean-field theory. One PEO-based LCBCs showed nanocylinder phase in the gray area. G, Gyroid; S, Sphere; C, Cylinder; L, Lamellar. Reproduced with permission from Matsen & Bates, 1996. Copyright 1996, ACS.

It has been reported that the LC mesophase preferentially orients with respect to the IMDS due to surface stabilization effects [Anthamatten et al., 1999]. As shown in Fig. 7.9, two possible mesogenic orientations relative to an interface (such as a substrate or the IMDS) are homeotropic or homogeneous, where the LC molecules is perpendicular or parallel to the interface, respectively. From the geometric point of view, homogenous anchoring are preferring when the mesogens are decoupled from the polymer backbone by relatively short spacers, whereas mesogens with a long spacer in LCBCs easily accept the homeotropic boundary condition [Osuji et al., 1999; 2000].

According to the extent of the block segregation in BCs, three regimes of microphase separation are defined as: weak ($\chi N \approx 10$), intermediate ($\chi N \approx 10{-}100$), and strong segregation ($\chi N \geq 100$) regimes. In LCBCs, χ is expected to be much larger than that in conventional amorphous BCs due to the mesophase formation. Such a large χ often leads to a much higher order–disorder transition (ODT). In some cases, it is difficult to measure the ODT of LCBCs because they may be decomposed before reaching their ODT. The large χ and high ODT make it possible to maintain microphase-separated structures of LCBCs above the clearing point (or in the isotropic phase) of the mesogenic block. These enable the phase diagram of LCBCs far different from that of amorphous BCs, in which the blocks might be miscible at a high temperature (>ODT).

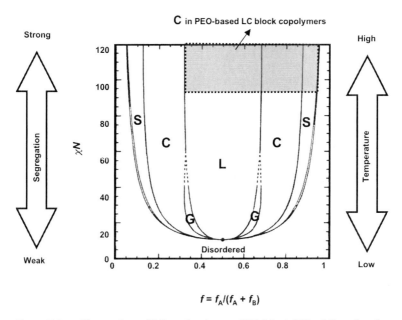

$$f = f_A/(f_A + f_B)$$

Figure 7.9 Illustration of LC anchoring on IMDS in LCBCs. LC molecules are homogeneous or hometropic alignment at the interface.

Owing to the structural asymmetricity, LCBCs often show an asymmetric phase diagram. For instance, a PEO-based diblock copolymer composed of a hydrophilic PEO block and a hydrophobic polymethacrylate containing an azobenzene moiety in its side group (Fig. 7.10), easily showed microphase separation with nanocylinder phase of PEO dispersed in LC matrix. Its nanocylinder phase range is given in the gray area of Fig. 7.8 [Yoon et al., 2008]. Recently, a nice bit of theoretical work (and some computational also) has been done to sketch out these differences [Lynd et al., 2008], which are in very good agreement with already reported experimental observations. However, it is difficult to give universal phase diagram like the amorphous AB-like diblock copolymer shown in Fig. 7.8 because of the complexity of LCBCs.

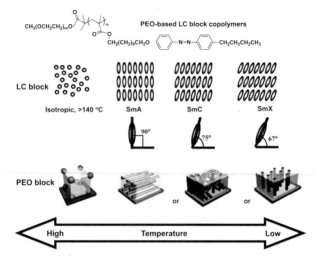

Figure 7.10 Microphase separation of PEO-based LCBCs. Reproduced with permission from John Wiley and Sons Publisher.

Heating LCBCs up to a temperature higher than the isotropic phase, their microphase separations often show changed morphologies passing the clearing point. Anthamatten & Hammond [1999] reported change in phase behaviors of polystyrene (PS)-based smectic LCBCs (with 50–58% volume-fraction of LC samples) from cylindrical or mixed-layered morphology to a lamellar morphology when heating them above the LC-to-isotropic temperature. Another example is one PEO-based LCBC (Fig. 7.10), showing PEO nanospheres when the azobenzene LC block is in its isotropic phase (>140°C). Cooling the PEO-based LCBC below its clearing point, the thermotropic LC block self-organized into smectic A (SmA), smectic C (SmC), and smectic X (SmX) [Tian et al., 2002]. Accordingly, order–order phase transition occurs, and the spherical PEO block changes into nanocylinders dispersed in continuous LC phases. The PEO nanocylinders are normal, parallel, or random patterning if the LCBCs are not underwent any special treatment.

7.3 Structures and Properties of LCBCs

Depending on the molecular structures (e.g., molecular weight of LCBCs, the kind of mesogenic units and LC phases), the relative volume fractions of the LC block, and the orientational interactions between different components, LCBCs can self-organize into diverse of nanostructures with additional functions. Confining the mesogenic block in nanospheres or nanocylinders upon microphase separation, the scattering of visible light can be eliminated when the LC blocks are in the minority phase. If the mesogenic blocks form the majority phase, supramolecular cooperative motion enables LCBCs to form macroscopically ordered nanostructures controlled by modulation of LC alignment. Obviously, the structure of LCBCs should exert great influence on their fascinating performance.

The photoresponsive azobenzene moiety may play the roles of both mesogens and photosensitive chromophores, when they are attached to the polymer main chain by a longer soft spacer. Both the photoisomerization and the LC-to-isotropic phase transition are involved in the process of microphase separation due to the immiscibility between azobenzene blocks and the non-azobenzene blocks. As shown in Fig. 7.11, all the processes in the microphase separation at different temperatures might be influenced by the photoresponse of azobenzene mesogens, and the microphase-separated nanostructures in azobenzene block copolymer films could have effect on the photochemical behavior of azobenzene blocks, vice versa. The photochemical control and supramolecular self-assembly make the azobenzene-containing LCBCs in solid state superior to that of their homopolymers or random copolymers.

7.3.1 Effect of Microphase Separation on LC Phases

Generally, LCBCs with LC blocks as the continuous phase show thermal properties and LC phase behavior similar to that of their LC homopolymers containing the same mesogenic unit. Recent work shows that confinement of the mesogenic blocks in the minority phase from microphase separation exerts great influence on the ordered structures, and LC performances of LCBCs might be modified comparing with their LC homopolymers. Fischer et al. [1994] studied the influence of the morphological structure on the LC behavior of PS-based LCBCs with cholesteryl groups as mesogens. Only a nematic phase was found for those samples with the LC block in separated spheres, whereas all samples with continuous LC matrix showed SmA phase, the same as that in the cholesteryl-containing homopolymers. Replacing the PS block with a poly(n-butyl methacrylate) in the cholesteryl LCBCs, the interaction of the non-LC block and the LC block was changed. Unexpectedly, the structural morphology effect on LC behavior of LCBCs showed similar results [Fischer et al., 1995]. Besides, the influence of LC content on the phase structures was also studied in a PS-based LCBC bearing bent-core mesogens [Tenneti et al., 2009].

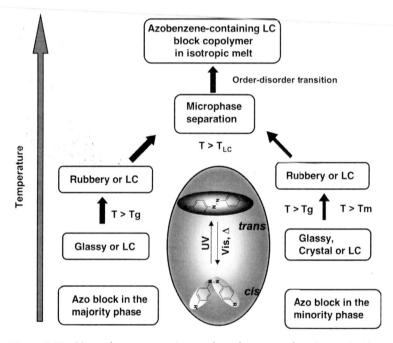

Figure 7.11 Microphase separation and azobenzene photoisomerization in bulk films of well-defined azobenzene LCBCs. T_g: glass transition temperature; T_{LC}: LC-isotropic phase transition temperature.

The LCBC microphase-separated morphologies show influence not only on LC phase structures but also on phase transition temperature. For instance, a polystyrene (PS)-based LCBC with mesogenic nanocylinders embedded within a PS matrix exhibited a clearing point, 22°C higher than that of an LCBC with a lamellar structure, even though the former sample has a slightly lower molecular weight than the latter one [Mao et al., 1997]. It was proposed that the cylindrical nanodomain structures in LCBCs might stabilize the smectic mesophase within it than the lamellar microphase morphologies.

The BC with a well-defined structure easily shows LC nature than copolymers with randomly structures when they possess a similarly low content of a mesogenic groups [Naka et al., 2009]. Thanks to the microphase-separated functions, the mesogenic block in BCs more likely segregates and assembles into an LC ordering in local areas, whereas the mesogenic groups in random copolymers are statistically distributed in a non-mesogenic matrix. Therefore, no microphase separation can be observed since the mesogenic block and other segments are miscible, leading to a disordered amorphous phase in films. As shown in Fig. 7.12, a random copolymer has a similar composition of mesogens (about 22 mol%) to that of a well–defined BC, and the former one did not exhibit an LC phase. Differently, a nematic LC phase was observed for the LCBC, which might be attributed to the effect of microphase separation. Such additional LC ordering makes the LCBCs more advantageous over the random copolymers. For instance, cooperative effect occurred in photoalignment of the segregated LC block, whereas it did not exist in random copolymer because this function was obstructed by the glassy PMMA. This resulted in a photoinduced birefringence in the LCBC film, larger than that of the random copolymer.

7.3.2 Effect of Non-LC Blocks

Various non-LC blocks often influence the performance of LCBCs, which can be occasionally non-neglected when the non-LC blocks constitute the majority phase. Under these circumstances, the non-LC blocks show larger effect on the performance of LCBCs comparing with the counteractive LC blocks in separated phases. For instance, the PEO block in LCBCs shown in Fig. 7.10 may supply hydrophilicity, ionic conductivity and crystallization [Yu et al., 2006a; 2006b]. Introduction of a PS block provides the designed LCBCs with a high glass-transition temperature (T_g) [Morikawa et al., 2007], and offers confinement effects on the photoinduced alignment of azobenzene LC side chains [Tong et al., 2004]. A polysiloxane backbone can greatly decrease T_gs of LCBCs and increase the LC temperature ranges

[Verploegen et al., 2008]. Putting rubbery poly(n-butyl acrylate) as middle block in ABA-type triblock copolymers, thermoplastic elastomers were obtained [Cui et al., 2004], which contrasts with conventional thermoplastic elastomers (such as styrene–butadiene–styrene triblock copolymer). Semi-fluorinated alkyl substituents of the mesogenic moieties may decrease the surface energy of LCBC films [Paik et al., 2007]. Using PMMA as one block, one LCBC film may show good optical transparency [Yu et al., 2008]. Importing regioregular poly(3-hexylthiophene) as one block provides LCBCs with conductivity [Han et al., 2010].

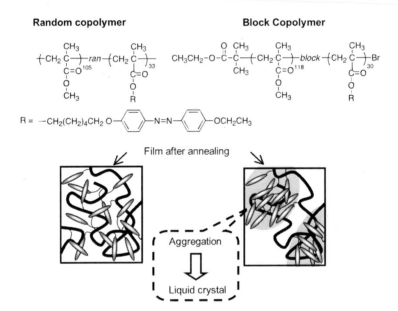

Figure 7.12 Schematic illustration of a well-defined BC and a random copolymer with a similar mesogenic content (about 22 mol%). As a result of the microphase separation, the BC shows LC phase, whereas the random copolymer is amorphous.

Upon microphase separation, the non-LC blocks also interfere with the LC self-assembly of the mesogenic block. Comparing with amorphous BCs, strong molecular interactions exist in LCBCs because of the inevitable self-assembly in LC domains and thermodynamically driven microphase separation. Investigation of LCBCs can undoubtedly fertilize the detailed illustration of microphase separation, and novel nanostructures and phase behaviors are expected in the process containing thermodynamic balances between microphase separation and LC elastic deformation. Recently, a novel worm-like nanostructure composed of mesogenic blocks was achieved in thin films of one smectic LCBC shown in Fig. 7.13. The LC texture of bar-coated films observed with polarizing optical microscope (POM) disappeared upon nanoscaled microphase separation because of the limitation of resolution of the used POM [Yu et al., 2007].

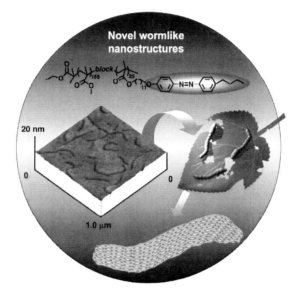

Figure 7.13 The non-LC block interferes the LC self-assembly of the mesogenic block in LCBCs, producing novel phase domains.

When the non-LC blocks form the continuous phase of LCBCs, they bring about confinement effect by dispersing the mesogenic domains in nanoscale. Although mesogens were constrained in the narrow space (like nanospheres and nanocylinders), MCM between photoresponsive LC moieties and other photoinert mesogenic groups was observed in Fig. 7.14 [Yu et al., 2007a]. Triggered by one beam of linearly polarized light, photoalignment of azobenzenes was first induced, followed with alignment of cyanobiphenyl mesogens coinciding with the ordered azobenzenes due to molecular cooperative motion. No obvious influence on the whole film was detected since the photoalignment occurred only in the separated phases. More interestingly, optical performance of the LCBC films was greatly improved because the molecular cooperative motion occurred only in nanoscale, eliminating the scattering of visible light. Accordingly, thick films (about 200 μm) with high transparency and low absorption based on the LCBC have been fabricated (Fig. 7.14), making it possible to record Bragg-type gratings for volume storage.

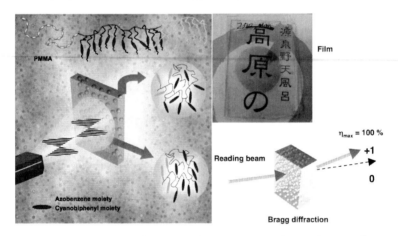

Figure 7.14 The non-LC blocks in the continuous phase of LCBCs, bring about confinement effect by dispersing the mesogenic domains in nanoscale, but molecular cooperative motion of mesogens still can be photoinduced.

7.3.3 Effect of LC Blocks

The LC blocks act as leading role in the scenario of LCBCs. LCPs with nematic, smectic, and cholesteric mesophases have been utilized to prepare LCBCs, introducing 1D, 2D, and helix ordering to the systems. The microphase separation of LCBCs with nanospheres or nanocylinders separately distributed in an LC matrix looks like the case of the LMWLC materials doped with dichroic dyes, in which the famous host–gust effect has been intensively studied for applications in optical devices. Just as that the guest dyes can be manipulated by changing orientation of the host LCs, the dispersed nanostructures in LCBCs also can be controlled with the help of the mesogenic matrix. These processes of LC-assisted supramolecular self-assembly in BCs have been regarded as supramolecular cooperative motion (SMCM,

Fig. 7.15), distinguishing from the molecular cooperative motion in LC mixtures. This interplay function, between the thermodynamical microphase separation and self-organized LC ordering in LCBCs, supplies unique method to macroscopically control regular nanostructures in LCBCs.

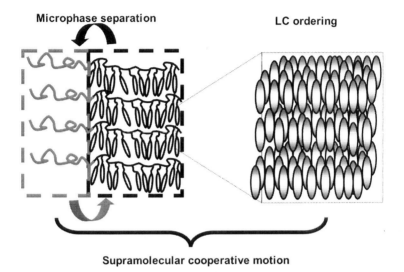

Figure 7.15 Scheme of supramolecular cooperative motion in well-defined LCBCs.

If the mesogenic block bearing photoresponsive groups (like azobenzene), such LCBCs are expected for potential applications in photonic devices, benefiting from both LC assembly and microphase separation. It was reported that enhancement of surface-relief gratings (SRGs) recorded in a PEO-based LCBC (Fig. 7.10) by thermally induced microphase separation [Yu et al., 2005b]. The diffraction efficiency of the gratings was also greatly increased by this post-treatment of annealing (Fig. 7.16).

PMMA-based LCBCs bearing azobenzene mesogenic blocks as the continuous phase showed similar capability of holographic recording [Yu et al., 2008], and both SRGs and refractive-index gratings (RIGs) were obtained. Decreasing the fraction of the LC block in PMMA-based LCBCs until the mesogens formed the minority phase, obvious confinement effect was observed in holographic recording. As shown in Fig. 7.17, only RIGs were obtained and SRGs were prohibited by the constraint since the micrometer-scaled mass transfer necessary for the surface-relief formation was forbidden by the glassy substrate of PMMA. However, these forbiddances of surface-relief formation evaded destroy of materials surface, which could benefit to holographic applications for volume storage [Häckel et al., 2005; 2007]. Furthermore, the microphase-separated nanostructures of LCBCs might provide ideal materials for photoresponsive artificial muscles [Li & Keller, 2006], which has been introduced in the Section 6.3 of Chapter 6.

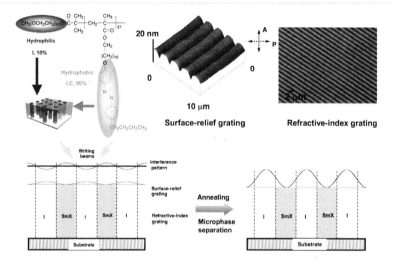

Figure 7.16 Holographic gratings recorded in azobenzene-containing LCBCs with well-defined structures and enhancement of gratings upon microphase separation. Reproduced with permission from John Wiley and Sons Publisher.

7.4 Control of Microphase Separation

Generally, the driving force for microphase separation of BCs is to achieve the required balance of minimizing the interfacial energy and maximizing the conformational entropy. However, thin films of microphase-separated BCs typically do not have long-range ordering, which limits their further utilization. LCPs and BCs are two types of ordered non-crystalline materials that can undergo self-assembly. From the viewpoint of molecular design, photoresponsive LCBCs integrate their unique characteristics of the two kinds of materials into one single system, which also bears photocontrol properties inherited from the chromophores. Recently, the supramolecular cooperative motion (SMCM) is regarded as one of the most effective approaches to control microphase-separated nanostructures of LCBCs [Yu et al., 2006a; 2006b], which leads to several newly developed approaches to macroscopically control the nanostructures in LCBCs.

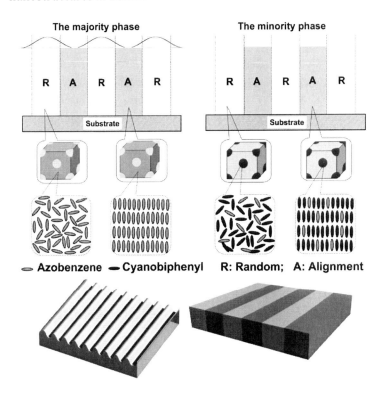

Figure 7.17 The LC blocks play an importance role in the photonic applications of LCBCs. Both surface-relief gratings and refractive-index gratings can be recorded when photoresponsive LC blocks forms the continuous phase, whereas only refractive-index gratings can be obtained if the mesogenic blocks are in the minority phase. (A, aligned, R, random). Reproduced with permission from Yu et al., 2008. Copyright 2008, ACS.

7.4.1 Thermal Annealing

Fabrication of large-area periodic nanostructures using supramolecular self-organizing systems is of great interest because of the simplicity and low cost of the process. Although macroscopically ordered microphase separation has been successfully obtained in amorphous BCs [Darling, 2007], both high reproducibility and mass production of such regularly ordered nanostructures through self-assembling nanofabrication processes still remain unsolved. The specially designed PEO-based LCBCs are good candidates to produce low-cost materials with self-assembled nanostructures, leading to industrial applications in future engineering plastics. As shown in Fig. 7.18, an almost perfect array of PEO nanocylinders with ordered alignment perpendicular to the substrates and with hexagonal packing was obtained upon thermal annealing, which is clearly observed by AFM, TEM, and FESEM, respectively [Komura & Iyoda, 2007]. Such regular periodic arrangements of nanocylinders were not limited to the film surface; cross-sectional images of TEM and AFM confirmed the formation of 3D arrays.

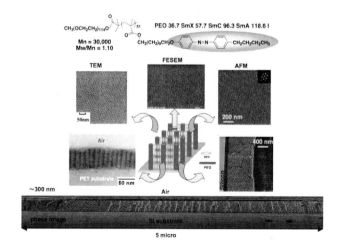

Figure 7.18 Perpendicular nanocylinder array of nanocylinders with the help of homeotropic LC alignment. Reproduced with permission from John Wiley and Sons Publisher.

Figure 7.19 shows the change in UV–Vis absorption spectra after the annealing process. Both a blue shift and great decrease in the absorption occur for the maximum peak due to the $\pi-\pi^*$ transition of azobenzenes. This indicates that the out-of-plane orientation of the smectic mesogens occurs upon annealing because of their layer structures. It has been proven that the 3D arranged nanostructures should be assisted by the out-of-plane orientation of azobenzene mesogens, which form the continuous phase in the LCBCs.

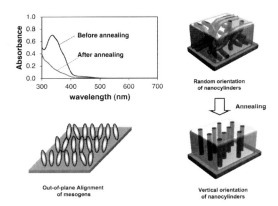

Figure 7.19 UV–V is absorption spectra of LCBC film and schemes of mesogens and nanocylinders after annealing.

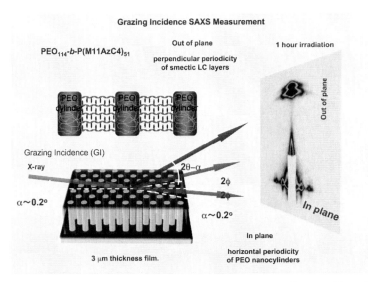

Figure 7.20 Measurement of grazing incidence SAXS of LCBC film after annealing.

In the TEM cross-sectional image, the smectic layer structures of azobenzene blocks were also observed normally to the PEO nanocylinders (parallel to the substrate). Such cooperative effect between hydrophilic PEO nanocylinders and the hydrophobic azobenzene mesogenic orientation is the result of supramolecular cooperative motion at a temperature higher than the smectic LC-isotropic phase transition temperature, which decreases the viscosity of LCBC films, and enables the interaction between the microphase separations with the smectic LC ordering to proceed completely. The measurement of grazing incidence SAXS also verifies that both the PEO nanocylinders and azobenzene mesogens are normal to the substrates upon thermal annealing [Komura et al., 2009] as shown in Fig. 7.20.

7.4.2 Mechanical Rubbing

Apart from the perpendicular patterning of densely packed nanostructures, parallel patterning in-plan of them was acquired by using a mechanical rubbing method. According to the SMCM in LCBCs, a long-range order of PEO nanocylinders can be obtained coinciding with the LC alignment direction, suggesting the application of other LC alignment techniques such as rubbing to control the microphase-separated nanostructures [Yu et al., 2006b]. As shown in Fig. 7.21, the LCBC film show a large anisotropy and a high order parameter upon rubbing treatment, indicating that the mesogens are aligned along the rubbing direction.

In Fig. 7.22, almost perfect alignment of PEO nanocylinders, coinciding with the LC orientation was clearly observed by AFM after rubbing and annealing treatments [Yu et al., 2007d; 2009b]. All the PEO nanocylinders were regularly oriented along the rubbing direction. This almost defect-free periodic nanocylinder array in the LCBC film can be acquired over arbitrarily large areas on the surface of rubbed polyimide films. Similar to other methods of controlling microphase-segregated domains in amorphous BC films, such as electric field, crystallization, controlled interfacial interaction, chemically or topologically patterned substrates, this rubbing technique exerted a forceful action on 3D array of nanostructures, as indicated from the cross-sectional images (Figs. 7.22 and 7.23).

Control of Microphase Separation | 235

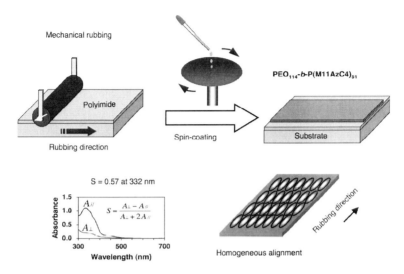

Figure 7.21 Treatment of PEO-based LCBC film with rubbed polyimide film.

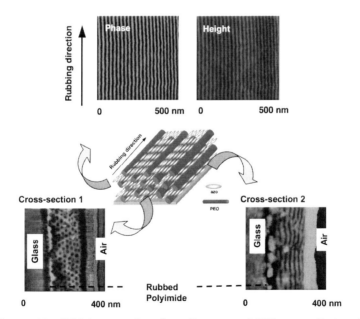

Figure 7.22 AFM images of perfect alignment of PEO nanocylinders in LCBC films, coinciding with the direction of LC orientation. Reproduced with permission from John Wiley and Sons Publisher.

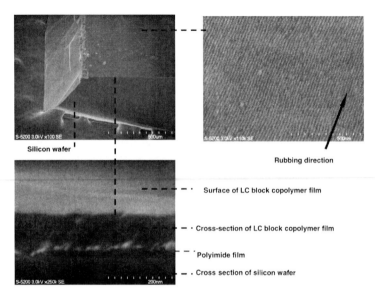

Figure 7.23 Field emission scanning electron microscopy (FESEM) pictures of rubbing-treated films one PEO-based LCBC.

7.4.3 Photoalignment

As described in Section 2.4 of Chapter 2, dust or static electricity might be produced in the rubbing process to control LC alignment, which could induce defects to the macroscopic array of microphase-separated nanostructures in LCBC films. Moreover, this contact method can only be applied on a flat surface; it shows no function on a curve one. Therefore, non-contact methods, such as light-controlled approaches have been explored [Yu et al., 2006a]. More perfect and simpler fabrication of microphase-separated nanostructures in thin films of azobenzene-containing LCBCs is expected by photoalignment method.

Upon irradiation of one beam of LPL, azobenzene chromophores are known to undergo photoalignment with transition moments almost perpendicular to the polarization direction by repetition of *trans–cis–trans* isomerization cycles, which has been discussed in Chapter 3. Such ordering can be transferred directly to other photo-inert mesogens by MCM coinciding with the ordered azobenzene moieties. By incorporating the photoalignment of azobenzenes into LCBCs with SMCM, the molecular ordering of the azobenzene can be transferred to a supramolecular level. Therefore, well-ordered nanostructures of the azobenzene-containing LCBC films might be obtained by the photocontrolled method. As shown in Fig. 7.24, one beam of LPL at 488 nm was used to manipulate the PEO nanocylinders self-assembled in an amphiphilic LCBC with well-defined structures.

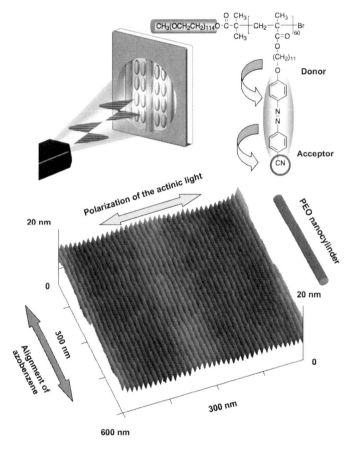

Figure 7.24 Photoalignment of nanocylinders in LCBC films, perpendicular to the polarization direction of the laser beam. Reproduced with permission from Yu et al., 2006a. Copyright 2006, ACS.

To enhance the absorption at 488 nm, a pseudostilbene-type azobenzene with a cyano group as an electron-acceptor substituent was used in preparation of the LCBC. Upon annealing without photoirradiation, a hexagonal packing of the PEO cylinders perpendicular to the glass substrates was obtained, due to the out-of-plane orientation of the smectic mesogens, which is similar to that in Fig. 7.18. The photoinduced homogeneous alignment of azobenzene moieties was carried out at room temperature, and then the anisotropic LCBC films were thermally annealed at a temperature just lower than the smectic LC-to-isotropic phase transition temperature. Perfect array of parallel PEO nanocylinders in-plan was achieved, aligned perpendicularly to the polarization direction of the laser beam due to SMCM. Thus, both parallel and perpendicular alignment of the nanocylinders is obtained in the same LCBC films, as shown in Fig. 7.25.

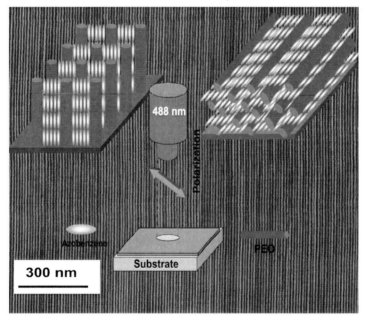

Figure 7.25 Both parallel and perpendicular alignment of the nanocylinders is obtained in the same LCBC films. Reproduced with permission from Yu et al., 2006a. Copyright 2006, ACS.

Recently, Morikawa et al. [2006] developed a control method of PEO nanocylinder by a periodic change in film thickness induced by mass transfer to form holographic gratings. In their preparation, the film thickness must be strictly modulated and the azobenzene-containing LCBC was mixed with a LMWLC (5CB) to assist to photoinduce a large mass transfer upon irradiation with an interference pattern, and 5CB was eliminated after grating formation. To simplify the process, they adopted one polarized beam to control the nanocylinders in PS-based azobenzene-containing LCBC films [Morikawa et al., 2007]. Defects appeared in the microphase-separated nanostructures, probably caused by incomplete microphase separation resulting from the high T_g of the PS block. Very recently, they in situ observed the photoreorientated process of microphase separation in LCBC films with time-resolved synchrotron X-ray scattering method [Nakano et al., 2012].

In non-doped films of PEO-based LCBCs with well-defined structures, macroscopically parallel patterning of PEO nanocylinders can be easily obtained in an arbitrary area by the simple and convenient photocontrol. Furthermore, the non-contact method might provide an opportunity to control nanostructures even on curved surfaces. Based on the principle of SMCM, the orientation of microphase-separated nanocylinders dispersed in mesogenic matrix should agree with the LC alignment direction. Moreover, the photocontrol of nanocylinders by SMCM was successfully obtained on nematic LCBCs, as shown in Fig. 7.26 [Yu et al., 2011]. Therefore, both in-plane and out-of-pane alignment of nanocylinders coinciding with azobenzene orientation might be precisely photocontrolled SMCM, which is expected to provide complicated nanotemplates for top–down nanofabrications such as lithography and beam processing.

For PEO-based LCBCs, the five states of nanocylinders (Fig. 7.27) can be easily achieved by modulation of the LC alignment of mesogenic blocks. When these LCBCs with PEO nanocylinders at room temperature were heated above the clearing point, the nanocylinders changed into nanospheres in the isotropic liquid-like LCBCs (Fig. 7.27f). Since the phase transition of mesogens can be reversible controlled, regular patterning of the nanostructures in LCBCs can be reversibly acquired, as shown in Fig. 7.27. In short, a smectic LC possessing a 2D order has confirmed to show the ability to control nanostructures of smectic LCBCs in a 3D order (Fig. 7.27). Recently, a nematic LC with a 1D order was also used to manipulate microphase-separated nanostructures in nematic LCBCs due to SMCM.

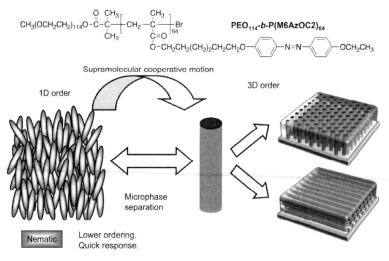

Figure 7.26 Photocontrol of nanocylinders in thin films of one nematic LCBC.

7.4.4 Electric and Magnetic Fields

Similar to the photocontrol method, other methods like electric and magnetic fields have been utilized to macroscopically control the microphase-separated nanostructures in LCBCs by combining with supramolecular cooperative motion (Fig. 7.28). More recently, an electrochemical method was developed to control alignment of PEO nanocylinders perpendicular to the substrate in PEO-based LCBC films [Kamata & Iyoda, 2008]. As shown in Fig. 7.28a, the LCBC films were prepared by spin coating on ITO glass kept at 50°C for two days. Using the ITO glass as a working electrode, a sandwich-type cell was assembled with a Teflon spacer and an injected KBr aqueous solution as electrolyte. Under the function of an electrolytic potential in the potentiostatic mode with Pt counter electrode and Ag/AgCl reference electrode, all the nanocylinders were oriented parallel to the electrolytic field as the lowest energy alignment, in spite of the microphase-separated state, parallel or random alignment of PEO nanocylinders. It was believed that ion diffusion locally induced in the vicinity of the electrode could allow the hydrophilic nanocylinders normal to the substrate, making it possible to manipulate the microphase-segregated microdomains.

Control of Microphase Separation | 241

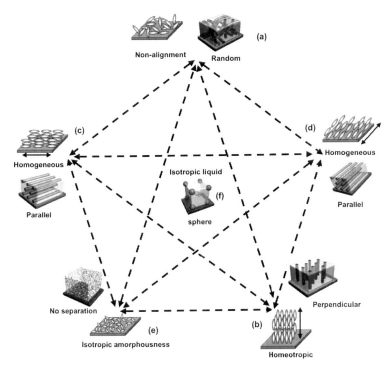

Figure 7.27 LC ordering helps self-assembly of PEO-based LCBCs. The arrows are directions of LC alignment. (a) Without any treatment, LCs are random and the nanocylinders are in non-alignment states. (b) Upon thermal annealing or function of external fields (electric and magnetic), LCs are in homeotopic alignment and nanocylinders are perpendicular to the substrate. (c) and (d) are homogeneously aligned LCs by rubbing treated substrate or photoalignment, and all the nanocylinders are parallel patterning. (e) Photoinduced LC-to-isotropic phase transition occurs when the azobenzene-containing LCBCs are irradiated with UV light, which causes disappearance of the nanocylinder phase at room temperature. (f) Heating the LCBCs above the clearing point, nanocylinders are changed into nanospheres dispersed in isotropic liquid-like LCBCs. Reproduced with permission from John Wiley and Sons Publisher.

For hydrogen-bonded LCBCs, an AC electric field was used to rapidly align the nanostructures at temperatures below the order–disorder transition but above T_g [Chao et al., 2004]. The LMW mesogens played an important role in controlling microphase separation. The fast orientation switching of the nanostructures was attributed to the dissociation of hydrogen bonds, which might be used to control nanostructures in supramolecular LCBCs.

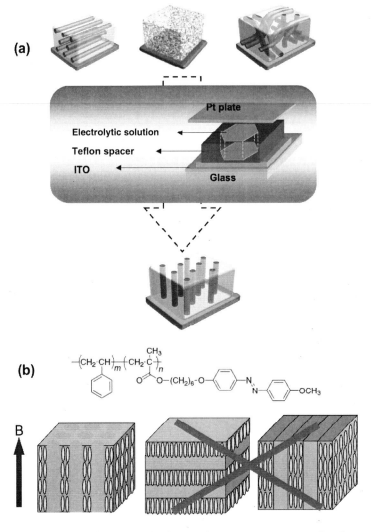

Figure 7.28 Control of nanocylinders in photoresponsive LCBCs by electric field and magnetic field.

Similar to the electric field, the magnetic field was used to manipulate the microphase separation in photoresponsive LCBCs.

The non-contact orientation method provides a higher degree of freedom for sample shapes than the mechanical orientation method. Furthermore, no danger presents, such as the dielectric breakdown that can be encountered in the electrical orientation approach. The uniform orientation of LCs could be obtained over the whole region of a sample, regardless of the macroscopic shape of the sample and the strength of the magnetic field [Tomikawa et al., 2005]. In Fig. 7.28b, hexagonally packed nanocylinders dispersed in mesogenic matrixes were aligned along the magnetic field upon annealing for a longer time (>2h) at a nematic LC phase. But the magnetic field showed no function on the lamellar-nanostructured LCBCs, possibly because ordered lamellar microdomains with a long correlation length were only rearranged very little [Osuji et al., 2004]. Although the LC was magnetically aligned in nanoscale layers, it showed no influence on the inverse continuous phase (Fig. 7.28b) [Hamley et al., 2004].

7.4.5 Other Methods

Besides, a shearing flow was explored to trigger the orientation of lamellar and cylindrical microdomains in LCBCs [Osuji et al., 1999; 2000]. Undoubtedly, other approaches for controlling microphase separation in films of amorphous BCs, such as solvent evaporation, film thickness, modified substrates, mixture with homopolymers, and roll casting can also be used with photoresponsive LCBCs.

It must be mentioned that the obtained nanostructures of photoresponsive LCBCs have a poor optical durability, which limits their wide application. To improve the stability of the obtained nanostructures, non-photosensitive aramid moieties were introduced as mesogens to form successive hydrogen bonds (H-bonds) in the matrix, as shown in Fig. 7.29. Furthermore, H-bonds might also be formed between the EO units and the aramid mesogens, and a plausible microphase-separated structures by supramolecular self-assembly is depicted in Fig. 7.29.

In summary, the photoresponsive LCBC films with well-ordered microphase separation show excellent reproducibility and mass production through molecular or supramolecular self-assembly, as shown in Fig. 7.30. This guarantees the nanotemplate-based nanofabrication process, and results in diverse self-assembled nanostructures, leading to widely industrial applications in plastics engineering.

Figure 7.29 AFM pictures of well-ordered microphase separation of thin films of one PEO-based LCBC with a large area (12,400 nm and 4100 nm).

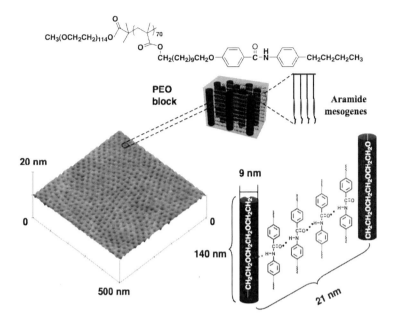

Figure 7.30 Stable nanocylinders in LCBCs with hydrogen-bonded mesogens.

7.5 Applications

7.5.1 Enhancement of Surface-Relief Gratings

As described in Section 5.5 of Chapter 5, holographic gratings have potential applications in information technology, which have been recorded on photoresponsive LCP films by utilizing their light-responsive properties. The diffraction efficiency is one of the most important parameters of holographic gratings. In amorphous polymer materials, a SRG contributes mainly to diffraction efficiency. It was reported that photoresponsive LCBCs were good candidates to control diffraction efficiency by enhancement of surface relief upon microphase separation [Yu et al., 2005; 2007c; 2007e; 2009a]. In Fig. 7.16, both a SRG and a RIG were recorded upon irradiation of an interference pattern, in which selective photoisomerization and the isotropic–to–LC phase transition were induced in the bright areas. The diffraction efficiency of the gratings depended strongly on the polarization of the reading beam because of the photoalignment of mesogens.

After grating formation, the surface-relief structure was clearly observed from the AFM images (Fig. 7.16), in which a sinusoidal curve was obtained. The fringe spacing of the surface-relief grating was 2.0 µm, identical to that of the RIG. Then microphase separation was induced in nanoscale by annealing the grating samples. As a result, the surface relief was increased to about 110 nm (18.3% of the film thickness), almost one order of magnitude larger than that before annealing. The peak-to-valley contrast became more explicit after annealing, due to the enhancement of the surface modulation. Furthermore, the sinusoidal shape of the surface profile became a little irregular, indicating that the LC alignment was disturbed upon microphase separation. Together with the enhancement of surface-relief structures, the diffraction efficiency increased to about 9.0%, almost two orders of magnitude larger than the diffraction efficiency before annealing. This increased diffraction efficiency may be ascribed mainly to the enhancement of surface modulation.

Comparing to other methods to control diffraction efficiency, such as gain effects, mechanical stretch, electrical switch, self-assembly, mixture with LC and cross-linking, the microphase-separation method had advantages of being simple and convenient. To precisely control diffraction efficiency of recorded gratings in photoresponsive LCBCs, the effect of recording time on grating formation and enhancement was studied systematically. The best enhancement effect was obtained at 10 s recording upon microphase separation. By adjusting the recording time, the diffraction efficiency was finely controlled from 0.13% to 10% [Yu et al., 2009].

7.5.2 Enhancement of Refractive-Index Modulation

By the cooperative effect between photoresponsive azobenzene moieties and photoinert groups, a small external stimulus can induce a large change in refractive index of the materials, which has been widely used in holographic recording. This is especially useful in photoresponsive LCBCs with azobenzene mesogens in the minority phase dispersed in glassy substrates. Then, the photoinduced mass transfer was greatly prohibited due to the microphase separation in grating recording, lack of surface-relief structures was observed [Breiner et al., 2007; Frenz et al., 2004]. Thus, refractive index modulation plays an important role in the grating formation in such photoresponsive LCBCs.

As shown in Fig. 7.17, holographic gratings were recorded in films of two PMMA-based LCBCs. One was a well-defined azobenzene-containing diblock copolymer, and the other sample was a diblock random copolymer. Here, the diblock random copolymer consisted of two blocks, in which one segment was PMMA and the other mesogenic block was statistically random. After grating formation, both films showed no formation of surface-relief structures, and only refractive-index modulations were obtained. Upon irradiation of two coherent laser beams, RIGs in the azobenzene-containing diblock copolymer was recorded by photoalignment of azobenzenes dispersed in phase-separated domains. In contrast, the photoalignment of the azobenzene was amplified by the photoinert cyanobiphenyl moieties as a result of the cooperative effect in the diblock random copolymer. This led to a similar refractive index modulation, although the azobenzene content was lower in the diblock random copolymer. The cooperative motion was confined within the nanoscale phase domains, unlike the case of random copolymers with statistically molecular structures.

7.5.3 Nanotemplates

Recently, fabrication of a well-arranged nanoparticle array by using nanostructured template films has become one of the important topics in nanotechnology. The size and the periodicity of nanoparticle arrays can be independently controlled by choosing appropriate templates, which is easily obtained by the macroscopic control of microphase separation in well-ordered nanotemplates of photoresponsive LCBCs. These nanotemplates of photoresponsive LCBCs are non-porous, different from the conventional porous nanotemplates. However, the PEO-based nanocylinders shown in Fig. 7.30 provide ultrahigh density nanoreactive space with high reliability, which can be filled with reactive compounds in the nanoscaled ethereal medium. By this way, diverse of nanomaterials can be prepared conveniently.

As shown in Fig. 7.31, a well-ordered array of Ag nanoparticles was successfully prepared over a large area on soft or hard substrates via selective Ag⁺ doping of the hydrophilic PEO domains in an LCBC film and an associated vacuum UV treatment to eliminate the LCBC templates to simultaneously reduce the Ag+. Obviously, the periodicity of the highly dense Ag nanoparticles can be precisely controlled by the nanotemplates of the LCBC films. The self-assembly from LCBC film templates also provides a good method to modify on the nanoscaled shape of various kinds of functional materials, such as electric conducting RuO_2, magnetic Fe, or organic conducting polymers [Li et al., 2007a].

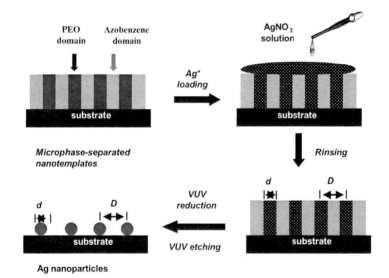

Figure 7.31 Fabrication of periodic array of Ag nanoparticles. Reprinted from Li et al. [2007a], with permission from John Wiley and Sons Publisher.

Figure 7.32 presents a PEO-based LCBC film, which exhibits well-ordered hydrophilic PEO nanocylinders with hexagonally packing embedded in an LC matrix. The anisotropic PEO nanocylinders can be used as ion conductive channels since PEO has been widely used as solid electrolyte. By incorporating $LiCF_3SO_3$ into the PEO nanocylinders, a supramolecularly complexed structure and anisotropic ion transportation were achieved based on the LCBC nanotemplates. Highly ordered ion-conducting PEO nanocylinder arrays with perpendicular orientation were formed by coordination between the lithium cations and the ether oxygen of the PEO blocks. The right of Fig. 7.32 shows a 3D illustration of the corresponding microphase-segregated structure, in which anisotropic ion transport is observed [Li et al., 2007b].

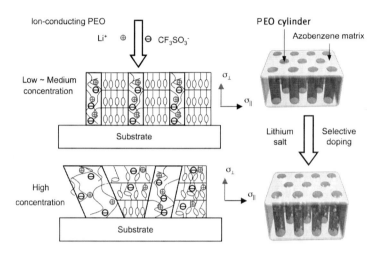

Figure 7.32 Anisotropic ionic conduction in nanochannels with nanotemplates of LCBC films. Reproduced with permission from Li et al., 2007b. Copyright 2007, ACS.

On the surface of PEO-based LCBC films, each hydrophilic PEO domain appears as a circular hollow surrounded by the hydrophobic LC matrix. These amphiphilic properties enable them to selectively absorb Au nanoparticles with hydrophilic or hydrophobic surface modifications. In the process of nanofabrication, the surface properties of the gold nanoparticles are a critical factor [Watanabe et al., 2007]. To extend site coverage further and favor high selectivity, the gold nanoparticles should be modified by additional functional ligands, which provides electrostatic and hydrogen bond interactions. As shown in Fig. 7.33, site-specific recognition of gold nanoparticles was obtained in the PEO nanocylinder domains or the continuous domains of LC blocks. Then the ordering of the gold nanoparticles was transferred to the substrate by a vacuum UV approach, which was effective in removing the templates without destroying the regularity of the assembled nanoparticles.

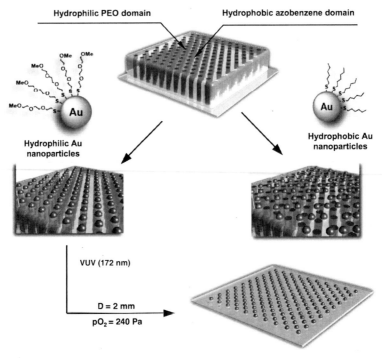

Figure 7.33 Selective absorption of Au nanoparticles on amphiphilic surfaces of LCBC films. Reprinted from Watanabe et al. [2007], with permission from John Wiley and Sons Publisher.

Besides, the sol–gel process can also be incorporated with LCBC film lithography. A hexagonally ordered SiO_2 nanorod array with mesochannels aligned along the longitudinal axes was obtained as shown in Fig. 7.34 [Chen et al., 2008]. By wet etching through LCBC thin films, nanodimple arrays with a hexagonal arrangement and a 24 nm center-to-center distance between nanodimples were successfully fabricated on the SiO_2 surface of a silicon wafer substrate. Perpendicularly oriented PEO cylinders in the LCBC film acted as nanochannels so that the etching reagent reaches the substrate surface [Watanabe et al., 2008]. The LCBC lithography with the smart film masks and wet etching process may be evaluated as a low-cost and mass-productive wet nanopatterning of silicon wafers, and may be applied to wide variety of substrates such as metals, semiconductors, glasses, and polymers by choosing appropriate etching reagents.

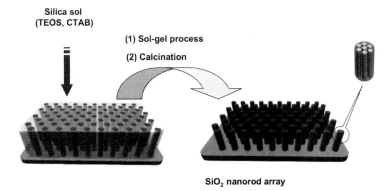

Figure 7.34 Potential representation of the preparation of SiO_2 nanorod arrays with mesochannels normal to the substrate, combining the sol–gel process with lithography based on photoresponsive LCBC films.

7.5.4 Microporous Structures

With an amphiphilic photoresponsive LCBC consisting of a flexible PEO segment as a hydrophilic part and poly(methacrylate) containing an azobenzene moiety in side chain as a hydrophobic one, well-arranged ellipsoidal micropores embedded in an LC matrix were fabricated by spin coating under a dry environment (Fig. 7.35) [Chen et al., 2010]. The formation process of the microporous films consisted of the following steps: Firstly, the LCBC was dissolved in THF, a good solvent for both segments. To avoid humid condition, a small amount of water was added to the THF solution. Secondly, the water-containing THF solutions were spin coated on clean glass slides. Upon spin coating, the evaporation of THF cooled the LCBC surfaces down and led to the formation of water droplets, which were stabilized by the amphiphilic LCBC and packed periodically. Finally, the regularly patterned microporous films were obtained after complete evaporation of water and THF.

With the help of small amount water, the obtained pore size was controlled in a range of dozens of microns [Chen et al., 2010]. Then, the influence of water content and rotational speed was studied in detail. It was found that regularly patterned microporous films could be prepared with certain water content, and the porous size could be easily tailored with changing the rotational speed. The obtained microporous structures showed good thermal stability below the LC-to-isotropic phase-transition temperature of the photoresponsive LCBC. Similarly, even the photoinduced LC-to-isotropic phase transition was induced upon UV irradiation at room temperature, the fabricated micropores showed no change in size upon photoirradiation. Although the mesogens in the LCBC films were randomly distributed, the LC property of self-organization might play an important role in the formation of linearly patterned ellipsoidal micropores embedded in a birefringent LC matrix with photoresponsive functions.

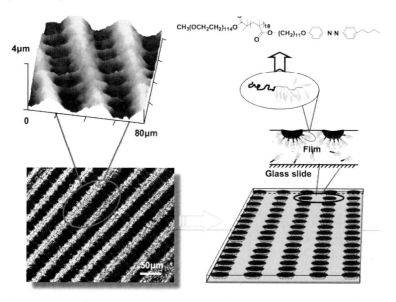

Figure 7.35 Fabrication of regularly patterned micropores with an amphiphilic photoresponsive LCBC by spin coating under a dry environment.

7.6 Outlooks

The microphase-separated nanostructures of well-defined photoresponsive LCBCs have fascinated one to understand the relationship between their ordered structures with photocontrollable properties of photoresponsive LCP blocks. One of the most exquisite advantages of introducing LCPs into well-defined photoresponsive LCBCs is precise photomanipulation of the supramolecularly self-organized nanostructures. With the development of information technology (IT), new waves are surging, driving such nanostructures leading to industrial applications as the future engineering plastics for optoelectronics and nanotechnology. Being expected as one of the powerful counterparts of top-down-type nanofabrication, the central focus should be placed on high reproducibility and mass production as well as precise manipulation of these ordered nanostructures. Although the research on photoresponsive LCBCs is still in a primary stage, many groups are involving in this novel field, which will improve our understanding of functional photoresponsive LCBCs and push ahead to find their diverse applications in optoelectronics, information storage, nanotechnology, as well as biotechnology.

References

Adams, J. and Gronski, W. (1989). LC side chain AB-block copolymers with an amorphous A-block and a liquid-crystalline B-block, *Macromol. Rapid Commun.* **10**, pp. 553–557.

Anthamatten, M., Zheng, W. and Hammond, P. (1999). A morphological study of well-defined smectic side-chain LC block copolymers, *Macromolecules* **32**, pp. 4838–4848.

Anthamatten, M. and Hammond, P. (1999). A SAXS dtudy of microstructure ordering transitions in liquid crystalline side-chain diblock copolymers, *Macromolecules* **32**, pp. 8066–8076.

Breiner, T., Kreger, K., Hagen, R., Hackel, M., Kador, L., Muller, A., Kramer, E. and Schmidt, H. (2007). Blends of poly(methacrylate) block copolymers with photoaddressable segments, *Macromolecules* **40**, pp. 2100–2108.

Chao, C., Li, X., Ober, C., Osuji, C. and Thomas, E. (2004). Orientational switching of mesogens and microdomains in hydrogen-bonded side-chain liquid-crystalline block copolymers using AC electric fields, *Adv. Funct. Mater.* **14**, pp. 364–370.

Chen, A., Komura, M., Kamata, K. and Iyoda, T. (2008). Highly ordered arrays of mesoporous silica nanorods with tunable aspect ratios from block copolymer thin films, *Adv. Mater.* **20**, pp. 763–767.

Chen, D., Liu, H., Kobayasi, T. and Yu, H. F. (2010). Fabrication of regularly –patterned microporous films by self-organization of an amphiphilic liquid-crystalline diblock copolymer under a dry environment, *Macromol. Mater. Eng.* **294**, pp. 26–31.

Cui, L., Tong, X., Yan, X., Liu, G. and Zhao, Y. (2004). Photoactive thermoplastic elastomers of azobenzene-containing triblock copolymers prepared through atom transfer radical polymerization, *Macromolecules* **37**, pp. 7097–7104.

Darling, S. (2007). Directing the self-assembly of block copolymers, *Prog. Polym. Sci.* **32**, pp. 1152–1204.

Finkelmann, H. and Bohnert, R. (1994). Liquid-crystalline side-chain AB block copolymers by direct anionic polymerization of a mesogenic methacrylate, *Macromol. Chem. Phys.* **195**, pp. 689–700.

Fischer, H., Poser, S., Arnold, M. and Frank, W. (1994). On the influence of the morphological structure on the liquid crystalline behavior of liquid crystalline side chain block copolymers, *Macromolecules* **27**, pp. 7133–7138.

Fischer, H., Poser, S. and Arnold, M. (1995). Liquid crystalline side group block copolymers with n-butyl methacrylate as an amorphous A-block: Synthesis and characterization, *Macromolecules* **28**, pp. 6957–6962.

Frenz, C., Fuchs, A., Schmidt, H. W., Theissen, U. and Haarer, D. (2004). Diblock copolymers with azobenzenebenzene side-groups and polystyrene matrix: Synthesis, characterization and photoaddressing, *Macromol. Chem. Phys.* **205**, pp. 1246–1258.

Gallot, B. (1996). Comb-like and block liquid crystalline polymers for biological applications, *Prog. Polym. Sci.* **21**, pp. 1035–1088.

Hamley, I., Castelletto, V., Lu, Z., Imrie, C., Itoh, T. and Al-Hussein, M. (2004). Interplay between smectic ordering and microphase separation in a series of side-group liquid-crystal block copolymers, *Macromolecules* **37**, pp. 4798–4807.

Häckel, M., Kador, L., Kropp, D., Frenz, C. and Schmidt, H. (2005). Holographic gratings in diblock copolymers with azobenzene and mesogenic side groups in the photoaddressable dispersed phase, *Adv. Funct. Mater.* **15**, pp. 1722–1727.

Häckel, M., Kador, L., Kropp, D. and Schmidt, H. (2007). Polymer blends with azobenzene-containing block copolymers as stable rewritable volume holographic media, *Adv. Mater.* **19**, pp. 227–231.

Kamata, K. and Iyoda, T. (2008). Alignment control and templating process in amphiphilic block copolymer thin film, *Research Report–NIFS–PROC Series* **70**, pp. 48–50.

Komura, M. and Iyoda, T. (2007). AFM cross-sectional imaging of perpendicularly oriented nanocylinder structures of microphase-separated block copolymer films by crystal–like cleavage, *Macromolecules* **40**, pp. 4106–4108.

Komura, M., Watanabe, K., Iyoda, T., Yamada, T., Yoshida, H. and Iwasaki, Y. (2009). Laboratory-GISAXS measurements of block copolymer films with highly ordered and normally oriented nanocylinders, *Chem. Lett.* **38**, pp. 408–409.

Lehmann, O., Forster, S. and Springer, J. (2000). Synthesis of new side-group liquid crystalline block copolymers by living anionic polymerization, *Macromol. Rapid Commun.* **21**, pp. 133–135.

Li, M. and Keller, P. (2006). Artificial muscles based on liquid crystal elastomers, *Phil. Trans. R. Soc. A* **364**, pp. 2763–2777.

Li, J., Kamata, K., Watanabe, S. and Iyoda, T. (2007a). Template- and vacuum-ultraviolet-assisted fabrication of a Ag-nanoparticle array on flexible and rigid substrates, *Adv. Mater.* **19**, pp. 1267–1271.

Li, J., Kamata, K., Komura, M., Yamada, T., Yoshida, H. and Iyoda, T. (2007b). Anisotropic ion conductivity in liquid crystalline diblock copolymer membranes with perpendicularly oriented PEO cylindrical domains, *Macromolecules* **40**, pp. 8125–8128.

Lynd, N. A., Hillmyer, M. A. and Matsen, M. W. (2008). Theory of polydisperse block copolymer melts: Beyond the Schulz–Zimm distribution, *Macromolecules* **41**, pp. 4531–4533.

Matsen, M. W. and Bates, F. S.(1996). Unifying weak- and strong-segregation block copolymer theories, *Macromolecules* **29**, pp. 1091–1098.

Makinen, R., Ruokolainen, J., Ikkala, O., Moel, K., Brinke, G., Odorico, W. and Stamm, M. (2000). Orientation of supramolecular self-organized polymeric nanostructures by oscillatory shear flow, *Macromolecules* **33**, pp. 3441–3446.

Matyjaszewski, K. and Xia, J. (2001). Atom transfer radical polymerization, *Chem. Rev.* **101**, pp. 2921–2990.

Mao, G., Wang, J., Clingman, S., Ober, C., Chen, J. and Thomas, E. (1997). Molecular design, synthesis, and characterization of liquid crystal-coil diblock copolymers with azobenzenebenzene side groups, *Macromolecules* **30**, pp. 2556–2567.

Morikawa, Y., Kondo, T., Nagano, S. and Seki, T. (2007). Photoinduced 3D ordering and patterning of microphase–separated nanostructure in polystyrene-based block copolymer, *Chem. Mater.* **19**, pp. 1540–1542.

Morikawa, Y., Nagano, S., Watanabe, K., Kamata, K., Iyoda, T. and Seki, T. (2006). Optical alignment and patterning of nanoscale microdomains in a block copolymer thin film, *Adv. Mater.* **18**, pp. 883–886.

Naka, Y., Yu, H. F., Shishido, A. and Ikeda, T. (2009). Photoresponsive and holographic behavior of an azobenzene-containing block copolymer and a random copolymer, *Mol. Cryst. Liq. Cryst.* **498**, pp. 118–130.

Nagano, S., Koizuka, Y., Murase, T., Sano, M., Shinohara, Y., Amemiya, Y. and Seki, T. (2002). Synergy effect on morphology switching: Real-time observation of photo-orientation of microphase separation in a block copolymer, *Angew. Chem. Int. Ed.* **51**, pp. 5884–5888.

Osuji, C., Zhang, Y., Mao, G., Ober, C. and Thomas, E. (1999). Transverse cylindrical microdomain orientation in an LC diblock copolymer under oscillatory shear, *Macromolecules* **32**, pp. 7703–7706.

Osuji, C., Chen, J., Mao, G., Ober, C. and Thomas, E. (2000). Understanding and controlling the morphology of styrene–isoprene side-group liquid crystalline diblock copolymers, *Polymer* **41**, pp. 8897–8907.

Osuji, C., Ferreira, P., Mao, G., Ober, C., Vander, J. and Thomas, E. (2004). Alignment of self-assembled hierarchical microstructure in liquid crystalline diblock copolymers using high magnetic fields, *Macromolecules* **37**, pp. 9903–9908.

Paik, M., Krishnan, S., You, F., Li, X., Hexemer, A., Ando, Y., Kang, S., Fischer, D., Kramer, E. and Ober, C. (2007). Surface organization, light-driven surface changes, and stability of semifluorinated azobenzene polymers, *Langmuir* **23**, pp. 5110–5119.

Serhatli, I. and Serhatli, M. (1998). Synthesis and characterization of amorphous-liquid crystalline poly(vinyl ether) block copolymers, *Turk. J. Chem.* **22**, pp. 279–287.

Sidorenko, A., Tokarev, I., Minko, S. and Stamm, M. (2003). Ordered reactive nanomembranes/ nanotemplates from thin films of block copolymer supramolecular assembly, *J. Am. Chem. Soc.* **125**, pp. 12211–12216.

Tenneti, K. K., Chen, X., Li, C. Y., Shen, Z., Wan, X., Fan, X., Zhou, Q. F., Rong, L. and Hsiao, B. S. (2009). Influence of LC content on the phase structures of side-chain liquid crystalline block copolymers with bent-core mesogens, *Macromolecules* **42**, pp. 3510–3517.

Tian, Y., Watanabe, K., Kong, X., Abe, J. and Iyoda, T. (2002). Synthesis, nanostructures, and functionality of amphiphilic liquid crystalline block copolymers with azobenzene moieties, *Macromolecules* **35**, pp. 3739–3747.

Tomikawa, N., Lu, Z., Itoh, T., Imre, C. T., Adachi, M., Tokita, M. and Watanabe, J. (2005). Orientation of microphase–segregated cylinders in liquid crystalline diblock copolymer by magnetic field, *Jpn. J. Appl. Phys.* **44**, pp. L711–L714.

Tong, X., Cui, L. and Zhao, Y. (2004). Confinement effects on photoalignment, photochemical phase transition and thermochromic behavior of liquid crystalline azobenzene-containing diblock copolymers, *Macromolecules* **37**, pp. 3101–3112.

Verploegen, E., Zhang, T., Jung, Y., Ross, C. and Hammond, P. (2008). Controlling the morphology of side chain liquid crystalline block copolymer thin films through variations in liquid crystalline content, *Nano Lett.* **8**, pp. 3434–3440.

Walther, M. and Finkelmann, H. (1996). Structure formation of liquid crystalline block copolymers, *Prog. Polym. Sci.* **21**, pp. 951–979.

Watanabe, S., Fujiwara, R., Hada, M., Okazaki, Y. and Iyoda, T. (2007). Site-specific recognition of nanophase-separated surfaces of amphiphilic block copolymers by hydrophilic and hydrophobic gold nanoparticles, *Angew. Chem. Int. Ed.* **46**, pp. 1120–1123.

Watanabe, R., Kamata, K. and Iyoda, T. (2008). Smart block copolymer masks with molecule-transport channels for total wet nanopatterning, *J. Mater. Chem.* **18**, pp. 5482–5491.

Yoshida, E. and Ohta, M. (2005). Preparation of micelles with azobenzene at their coronas or cores from nonamphiphilic diblock copolymers, *Colloid Polym. Sci.* **283**, pp. 521–531.

Yoon, J., Jung, S., Ahn, B., Heo, K., Jin, S., Iyoda, T., Yoshida, H. and Ree, M. (2008). Order–order and order–disorder transitions in thin films of an amphiphilic liquid crystalline diblock copolymer, *J. Phys. Chem. B* **112**, pp. 8486–8495.

Yu, H. F., Shishido, A., Ikeda, T. and Iyoda, T. (2005a). Novel amphiphilic diblock and triblock liquid-crystalline copolymers with well-defined structures prepared by atom transfer radical polymerization, *Macromol. Rapid Commun.* **26**, pp. 1594–1598.

Yu, H. F., Okano, K., Shishido, A., Ikeda, T., Kamata, K., Komura, M. and Iyoda, T. (2005b). Enhancement of surface-relief gratings recorded in amphiphilic liquid-crystalline diblock copolymer by nanoscale phase separation, *Adv. Mater.* **17**, pp. 2184–2188.

Yu, H. F., Shishido, A., Ikeda, T. and Iyoda, T. (2005c). Photoresponsive behavior and photochemical phase transition of amphiphilic diblock liquid-crystalline copolymer, *Mol. Cryst. Liq. Cryst.* **443**, pp. 191–199.

Yu, H. F., Iyoda, T. and Ikeda, T. (2006a). Photoinduced alignment of nanocylinders by supramolecular cooperative motions, *J. Am. Chem. Soc.* **128**, pp. 11010–11011.

Yu, H. F., Li, J., Ikeda, T. and Iyoda, T. (2006b). Macroscopic parallel nanocylinder array fabrication using a simple rubbing technique, *Adv. Mater.* **18**, pp. 2213–2215.

Yu, H. F., Asaoka, A., Shishido, A., Iyoda, T. and Ikeda, T. (2007a). Photoinduced nanoscale cooperative motion in a novel well-defined triblock copolymer, *Small* **3**, pp. 768–771.

Yu, H. F., Shishido, A., Iyoda, T. and Ikeda, T. (2007b). Novel wormlike nanostructure self-assembled in a well-defined liquid-crystalline diblock copolymer with azobenzene moieties, *Macromol. Rapid Commun.* **28**, pp. 927–931.

Yu, H. F., Shishido, A., Ikeda, T. and Iyoda, T. (2007c). Photoinduced alignment and multi-processes of refractive-index gratings in pre-irradiated films of an azobenzene-containing liquid-crystalline polymer, *Mol. Cryst. Liq. Cryst.* **470**, pp. 71–81.

Yu, H. F., Li, J., Shishido, A., Iyoda, T. and Ikeda, T. (2007d). Control of regular nanostructures self-assembled in an amphiphilic diblock liquid-crystalline copolymer, *Mol. Cryst. Liq. Cryst.* **478**, pp. 271–281.

Yu, H. F., Shishido, A., Li, J., Iyoda, T. and Ikeda, T. (2007e). Stable macroscopic nanocylinder arrays in an amphiphilic diblock liquid-crystalline copolymer with successive hydrogen bonds, *J. Mater. Chem.* **17**, pp. 3485–3488.

Yu, H. F., Shishido, A. and Ikeda, T. (2008a). Subwavelength modulation of surface relief and refractive index in pre-irradiated liquid-crystalline polymer films, *Appl. Phys. Lett.* **92**, pp. 103117(1–3).

Yu, H. F., Naka, Y., Shishido, A. and Ikeda, T. (2008b). Well-defined liquid-crystalline diblock copolymers with an azobenzene moiety: Synthesis, photoinduced alignment and their holographic properties, *Macromolecules* **41**, pp. 7959–7966.

Yu, H. F., Shishido, A., Iyoda, T. and Ikeda, T. (2009a). Effect of recording time on grating formation and enhancement in an amphiphilic diblock liquid-crystalline copolymer, *Mol. Cryst. Liq. Cryst.* **498**, pp. 29–39.

Yu, H. F. and Kobayasi, T. (2009b). Fabrication of stable nanocylinder arrays in highly birefringent films of an amphiphilic liquid-crystalline diblock copolymer, *ACS Appl. Mater. Interfaces* **1**, pp. 2755–2762.

Yu, H. F., Kobayasi, T. and Ge, Z. (2009c). Precise control of photoinduced birefringence in azobenzene-containing liquid-crystalline polymers by post functionalization, *Macromol. Rapid Commun.* **30**, pp. 1725–1730.

Yu, H. F., Kobayasi, T. and Hu, G. (2011). Macroscopic control of microphase separation in an amphiphilic nematic liquid-crystalline diblock copolymer, *Polymer* **52**, pp. 1154–1161.

Zhao, Y., Qi, B., Tong, X. and Zhao, Y. (2008). Synthesis of double side-chain liquid crystalline block copolymers using RAFT polymerization and the orientational cooperative effect, *Macromolecules* **41**, pp. 3823–3831.

Index

absorption 66, 69, 70, 74, 123, 233, 237
actinic light 52
   incident 118
   polarized 143
AFM image *see* atom force microscopy image
alignment 45, 51, 60, 61, 70, 71, 81, 82, 84, 87, 94, 105, 112, 142, 143, 147, 150, 180, 187
   homeotropic 221
   homogenous 86, 189
   in-plane 144
   lowest energy 240
   macroscopic 44
   out-of-plane 239
   photocontrolled 11, 133
   photoinduced 83, 176, 225, 258
   planar 107, 108
   random 108, 240
   uniaxial 68
   unidirectional 53, 100
   vertical 84, 107, 108
alignment direction 80, 87, 88, 121, 151, 181, 187, 188, 190, 202–204
alignment layers 41–45, 50, 51, 53, 56–58, 60, 68, 150, 185, 187
amorphous materials 85, 171
anisotropic change 176
anisotropic contractions 194
anisotropic control 150
anisotropic crosslinking 76
anisotropic gel 117
anisotropic geometry 31
anisotropic photocrosslinking 53, 55, 147
anisotropy 36, 56, 61, 64, 68, 76, 105, 211
   fluid 13
   induced 145
   structural 165
   uniaxial 38
annealing 56, 73, 74, 78, 79, 89, 229, 233, 237, 243, 246
applications 9, 11, 41, 63, 87, 89, 91, 100, 101, 124, 137, 139, 141, 160, 161, 194, 245
   industrial 232, 244, 253
   optical 120, 149
   technological 8
atom force microscopy image (AFM image) 43, 126, 157, 185, 232, 234, 254
atom transfer radical polymerization (ATRP) 214, 215, 254, 255, 257
ATRP *see* atom transfer radical polymerization
azobenzene chromophores 106, 140, 182, 207, 210, 236
azobenzene derivatives 66, 115, 129, 169, 183
azobenzene dyes 51–53
azobenzene mesogens 71, 80, 140, 142, 159, 161, 182, 186–188, 202, 204, 223, 233, 234, 246
azobenzene moieties 69, 70, 79–81, 133, 137, 139, 140, 143–146, 153, 154, 156, 159, 168, 172, 185–187, 192, 256, 258

azobenzene molecules 52, 69, 86,
  92, 93, 95, 99, 112, 146,
  181, 189, 208
azobenzenes 51–53, 65, 66, 69,
  70, 79, 80, 91–99, 101,
  102, 111–113, 117, 125,
  129, 130, 136–139, 144,
  145, 155, 181–184, 193
  achiral 105
  donor–acceptor 140
  guest 106, 111
  low-molecular-weight 217
  pseudostilbene-type 154, 237
  side-chain 139

back-isomerization 95, 102, 139,
  140, 142, 146, 182, 204
BCs see block copolymers
beam 79, 83, 124, 154, 156, 159,
  246
  actinic 124
  argon-ion 50
  coherent 153
  electron 206
  output 156
  polarized 239
  zeroth-order 159
birefringence 3, 5, 32, 39, 71, 78,
  84, 89, 93, 96, 136, 142,
  145–147, 149, 171, 172
block copolymers (BCs) 15, 157,
  158, 169, 179, 211, 212,
  217, 220, 225, 228, 230,
  253–256
blocks 161, 179, 184, 212, 218,
  221, 226, 247
  constituent 86
  hard 157
  isoprene 216
  non-azobenzene 223
  soft 157
bulk films 86, 224

cells 52, 57, 102, 112, 124, 185

glass 185
sandwich-type 240
chiral dopant 21, 99, 105, 113
chirality 17, 114, 116
cholesteric LCs (CLCs) 9, 16, 20,
  103–109, 117
chromophores 55, 91, 95, 106,
  113, 145, 176, 231
circularly polarized light (CPL)
  20, 70, 99, 163
CLCs see cholesteric LCs
CLCs
  photoresponsive 108
composite films 100–103, 129
contraction 181, 184, 191, 203,
  204
  induced 183
  local 202
  photoinduced 183, 193
CPL see circularly polarized light
crosslinkers 35, 176, 178, 179,
  185, 191, 193, 194, 202
crosslinking 158, 175, 178, 180,
  181, 195, 198
  chemical 179, 209
  covalent 182
  physical 179, 180

defects 19, 20, 24, 44, 236, 239
deformation 24, 85, 165, 186, 207
  large 176, 180
  mechanical 199
  photoinduced 125, 166, 181,
    182, 186, 191, 192
  photoinduced macroscopic
    166
  uniaxial 182
diamagnetism 38, 39, 45
diarylethenes 67, 98, 99, 106
diblock copolymers 247, 254, 256
  amorphous AB-like 221
  amphiphilic LC 161
  photoresponsive AB-type 213

differential scanning calorimetry (DSC) 21–24, 27
diffraction efficiency 152–155, 158–160, 172, 229, 245, 246
dipole–dipole interactions 116, 123, 143
dipole–quadruple interactions 105
double-twist cylinders 19, 20
double-twist structures 19, 20
DSC *see* differential scanning calorimetry
dyes 98, 121–123, 135
   anthraquinone 120
   dichroic 160, 228
   light-sensitive 120
   organic 141

elastic constants 6, 36, 37, 128
elastic deformation 219, 227
elastomers 11, 176, 182, 208, 253
   amorphous photoresponsive 181
   crosslinked 175
   rubbery 180
   triblock 179
electric field 20, 38, 45, 63, 101, 102, 108, 111, 112, 118, 119, 190, 191, 217, 234, 241, 242, 253
   external 100
   high-frequency 101
   low-frequency 101
   opposite 112
energy 25, 50
   free 118
   interfacial 218, 230
   mechanical 86, 181
expansion 181, 184, 197
   local 203
   photoinduced 125, 126, 193

fabrication 46, 56, 117, 150, 166, 206, 209, 236, 247, 252, 253
FESEM *see* field emission scanning electron microscopy
field emission scanning electron microscopy (FESEM) 232, 236
films 45–47, 53, 58, 153, 154, 158, 163, 165, 166, 168, 190, 191, 194, 200, 204–206, 225, 227, 247
   adhesive-free bilayer 206
   alignment-layer 68
   amorphous BC 234, 243
   anisotropic 76
   cholesteric LC 126, 128
   contracted 168
   flexible plastic 203
   free-standing 163
   hybrid 164–168
   irradiated 71
   nanostructured template 247
   non-doped 239
   rubbing-treated 236
   transparent 70, 158
   unstretched 203
function 12, 44, 45, 50, 54, 57, 61, 102, 108, 165, 223, 225, 236, 240, 241, 243
   interplay 229
   photocontrollable 116
   static 107

gelators 116, 117
glass slides 43, 46, 166, 185
   clean 251
glass substrates 44, 51, 58, 73, 238
   coated 151
gold nanoparticles 249
gratings 155, 156, 171, 229, 230, 246, 254
   Bragg 159

flat-structured 156
mechanically tunable 157, 158
refractive-index 153, 155, 229, 231, 258
groups 31–34, 67, 72–74, 89, 90, 148, 171, 216, 221, 253–255
anthracene 55
azotolane 146, 147
bipolar carrier-transporting 150
chiral 105, 106
cholesterol 73, 114
cholesteryl 216, 223
chromophore 91
cinnamate 54, 55, 147
cyano 237
cyanobiphenyl 144
donor–acceptor 146
hydrophilic oligooxyethylene 161
hydroxyl 216
light-responsive 137
mathacryloyl 177
phenyl 2
pheylcyclohexane 31
photoinert 246
photoisomerizable azobenzene 137, 139
photoresponsive 79, 137, 216, 229
pyridyl 198
side-chain 73
siloxane 154
terminal 35, 160
vinyl 177

helical pitch 103, 105, 106, 108
helical structures 104, 106, 108
 induced 106
 periodical 108
helical twisting power (HTP) 99, 105–108, 128, 130

HTP *see* helical twisting power
human muscles 191, 194

irradiation 69–71, 76, 79–83, 88, 90, 93, 95, 98–103, 143, 163, 165, 182, 183, 189–191, 203–206, 245
 alternating 156, 163
 electric-beam 178
 ion-beam 50
 laser 101
 pulse 112, 141
 slantwise 82
isomerization 66, 69, 70, 80, 93, 97, 99, 106, 139, 144, 168, 180, 182
isomers 69, 72, 74, 75, 92, 94, 95, 97, 98, 106
isotropic liquid 1, 13, 18, 41
isotropic phase 3, 5, 12, 20, 24, 28, 32–35, 84, 93, 95, 96, 141, 155, 221, 222

laser beam 71, 119, 122, 191, 192, 197, 199, 237, 238
 actinic 71, 192, 201
 coherent 44, 247
 polarized 118, 206
layers 100, 161
 adhesion 204
 emissive 149
 gold 46, 47
 inorganic 49
 photoconductive 100
 ultrathin 45
LC alignment 41–51, 53–58, 70, 78, 83–85, 87, 100, 118, 120, 122, 125, 167, 177, 241, 246
 homeotropic 45, 48, 49, 86, 232
 multi-domain 56
 photoinduced 71

LCBCs *see* LC block copolymers
LCBCs
   amphiphilic 214, 237, 251
   azobenzene 224
   cholesteryl 224
   hydrogen-bonded 241
   lamellar-nanostructured 243
   microphase-separated nanostructures 230, 231
LC block copolymers (LCBCs) 12, 86, 87, 161, 211, 212, 214, 216–231, 233, 234, 236, 237, 239–241, 243, 250, 251
LC blocks 179, 211, 222–224, 228, 231, 249
LC cells 51, 57, 58, 100, 103, 156, 183
   homeotropic 118
   nematic 118
   planar 108
   thin 111
LC directors 85, 118
LC displays (LCDs) 1, 8, 10, 15, 36, 41, 42, 55, 61, 64, 149
LC droplets 100, 102, 103
LCDs *see* LC displays
LCDs
   full-color 104, 108
   soft 44
LCE films 184, 186, 187, 189, 190, 192, 193, 195, 199, 201, 202, 207
   bonded 199
   crosslinked 198
   ferroelectric 191
   freestanding 176, 185, 186, 198
   homeotropic-alignment 188
   hybrid 190
   hydrogen-bonded 198, 199
   polydomain 187, 188
   self-assembled 199
   uniaxial 200

LC elastomers (LCEs) 85, 86, 166, 175–185, 189, 191–197, 199, 201–206
LCEs *see* LC elastomers
LCEs
   anisotropic 178
   azobenzene side-on 183
   copolymer-based 184
   dye-doped 199, 200
   homogeneously aligned 179, 190, 193
   miniature 194
   photoresponsive 86, 184, 196, 199
   photosensitive 201
LC materials 2, 4, 6, 10–13, 21–24, 28, 31, 36–41, 80, 82–84, 140, 145, 149, 153, 176
LC matrix 221, 228, 248, 251
   continuous 224
   hydrophobic 249
LC mixtures 94, 95, 179, 193, 229
LC molecules 33–35, 39, 41, 42, 44–47, 50, 52–54, 63, 64, 66, 67, 84–86, 94, 118, 122, 123, 151, 153, 220, 221
   photoresponsive 87
   rod-like 32, 68
LC ordering 87, 180, 181, 211, 218, 225, 241
LC orientation 234, 235
LCP films 70–72, 74, 141, 142, 156, 157, 160
   photocrosslinkable 76
   side-chain type 166
   spin-coated 70
LC phases 6, 8, 12, 13, 15–19, 21–24, 26, 27, 31, 33–36, 66, 68, 91–94, 96, 137, 144, 225, 226
   chiral 17, 19
   cholesteric 105
   discotic 21
   nematic 135, 137, 140

smectic 70, 144
LCP microparticles 164–166
LC polymers (LCPs) 15, 17, 70,
  72–76, 79, 82, 133,
  136–147, 150–154, 158,
  160, 161, 175, 178, 228,
  230
LCPs *see* LC polymers
LCPs
  aligned 151, 178
  brushlike 163
  end-on 15
  main-chain 15
  photocrosslinkable 78, 79,
    161
  self-organized 180
  side-chain 15, 101, 133, 143,
    144, 154
  side-on 15
  smectic 74
LCs *see* liquid crystals
LCs
  amphotropic 14
  antiferroelectric 113
  discotic 10, 21
  enantiotropic 23
  homogeneously aligned 241
  light-responsive 83
  lyotropic 14
  monotropic 23
  non-chiral 17
  photoinert 64
  polymer-dispersed 100
  polymer network 100
  polymer-stabilized 100
  rod-like 31
  thermotropic 12–14, 16
LC samples 24, 26, 27, 39, 95, 222
LC systems 63, 67, 95, 105
  hybrid 168
  light-responsive 91
  transparent 118
LC temperature 76–78, 93, 185,
  225

LC texture 125
light intensity 118, 119, 123, 127,
  172
light irradiation 68, 69, 74, 108,
  109, 127, 163, 196, 218
light source 24, 86, 165, 186, 189,
  190, 193, 194
  actinic 86, 188, 191, 202
  excitation 193
linear LCPs 176, 178, 180
linearly polarized light (LPL) 24,
  52, 53, 60, 61, 68, 69, 72,
  74, 75, 84, 89, 119, 120,
  144, 161, 187, 188, 227
liquid crystal (LC) 1, 2, 4, 5–8,
  10–13, 19–21, 28, 29, 45,
  59–61, 63, 64, 66–68,
  82–92, 94, 110–114, 118,
  122–126, 128–131,
  154–156, 167–173,
  178–180, 206–210, 226,
  227, 253–258
LMWLCs *see* low-molecular-weight
  liquid crystals
low-molecular-weight liquid
  crystals (LMWLCs) 91–94,
  96–98, 100–104, 106–108,
  110, 112, 114, 116–118,
  120–128, 130, 136, 140,
  176, 239
LPL *see* linearly polarized light

macromolecules 15, 29, 60,
  88–90, 168–172, 175, 181,
  208, 211, 253–256, 258
magnetic fields 7, 39, 41, 45, 184,
  211, 240, 242, 243, 256
MCM *see* molecular cooperative
  motion
mechanism 43, 50, 51, 95, 105,
  139, 141, 202
microgroove 44
photodriven 191

mesogenic blocks 211, 221, 223, 225–227, 229, 231, 239, 247
mesogenic groups 181, 225
  hydrogen-bonding 217
  photoinert 133, 153, 227
mesogens 15, 33, 34, 36–38, 78, 79, 136–139, 142, 143, 146, 147, 155, 156, 176, 180, 181, 184, 185, 190, 191, 220, 223, 227–229
  aligned 184, 192
  aramid 243
  azotolane 146–148
  bent-core 224, 256
  biphenyl 56
  birefringent 133
  homeotropic alignment 189
  hybrid-aligned 189, 190
  hydrogen-bonded 245
  monodomain-aligned 185
  non-photoactive 143
  ordered 177
  photochromic 184
  photoinactive 145
  photoinert 159, 236
  rod-like 79
  smectic 233, 238
mesophases 12, 15, 20, 24, 25, 28
microactuators 194
microphase separation 100, 102, 150, 157, 212, 220, 222–227, 229–231, 233–235, 237, 239, 241–243, 245–247, 254, 256
microscope 185
  optical 93, 96, 142, 185, 227
  regular 24
moiety 2, 65, 137, 150
  acceptor 140
  amine 150, 170, 171
  benzophenone 57
  carbazole 151
  cyano 2
  cyanophenyl benzoate 136
  ester 123, 124, 131
  hydrophobic azotolane 161
  mesogenic 82, 137, 226
  non-photosensitive aramid 243
  photoactive 138
  photoresponsive LC 227
  urea 114
molecular cooperative motion (MCM) 68, 76, 93, 94, 121, 125, 143, 145, 146, 211, 227–229
molecular motors 107, 125, 126, 128
molecular orientation 36, 41, 73, 83, 89, 161, 170, 190, 191
molecular weight 4, 12, 15, 122, 212, 222, 225
molecules 12, 15–17, 21, 28, 33, 36, 43, 68, 69, 81, 82, 99, 106, 123, 140, 197
  active 53
  amphiphilic 13
  amphotropic 15
  banana-shaped 18
  motor 125
  non-chiral 17
  organic 31
  pentadecylphenol 217
  photoresponsive 87, 101, 113
  rod-like 21, 94
monomers 34, 35, 178, 179, 191, 193, 202, 213, 214
  acrylate 34, 102
  azobenzene 183
  methacrylate 34, 35
  photoresponsive mesogenic 218
  polymerizable 214

nanocylinders 90, 217, 223, 227, 228, 232, 233, 237–242, 257

hydrophilic 240
mesogenic 224
microphase-separated 239
nanostructures 87, 211, 217, 223, 227, 231, 234, 239, 241–243, 253, 256, 258
nematic phase 16, 24, 25, 27, 99, 140, 223

optical anisotropy 24, 39, 61, 78, 145
optical fields 63, 119

PEO nanocylinders *see* poly(ethylene oxide) nanocylinders
phases 12, 13, 18, 20, 21, 23, 27, 113, 114, 127, 130, 149, 151, 153, 157, 223, 225, 226, 228
 birefringent 118
 columnar discotic 21
 continuous 223, 229, 231, 233, 243
 crystalline 24
 disordered amorphous 225
 minority 157, 158, 223, 229, 231, 246
phase transitions 7, 13, 22, 40, 92, 95, 103, 143, 163, 180, 181, 239
 isotropic-to-LC 80, 139
 isotropic–to–LC 245
 LC-isotropic 181
 nematic LC-to-isotropic 8, 96, 135, 140, 141
 nematic-to-cholesteric 99
 nematic-to-isotropic 98
 photochemical 95, 140
 photoinduced LC-to-isotropic 252
 photoinduced nematic-to-isotropic 154

phototriggered sol–gel 116
pohotochemical 97
photoalignment 50, 51, 53, 55, 57, 58, 69–75, 77, 79, 81, 143–145, 225, 227, 236, 241, 246, 247, 256
photoalignment layers 53, 55, 56, 58, 61
photochemical processes 119, 120, 122
photochemical reactions 51, 56, 64, 65, 67, 79, 95, 137
photochromic molecules 11, 63, 67–69, 91, 94, 97–99, 105, 125, 136
photoinduced birefringence 78, 147, 148, 173, 225
photoinduced change 84, 95, 105, 106, 140, 146, 147, 171
photoinduced phase transition 11, 80, 91, 93, 95–101, 138, 139, 146, 155, 163, 180
photoinduced reorientation 74, 89, 122–124, 131, 171, 192
photoirradiation 51, 57, 58, 83, 96–98, 102, 105, 108, 111, 112, 114, 135, 139, 141, 165, 166, 204, 208
photoisomerization 51, 53, 65–67, 79, 81, 91, 92, 95, 97, 98, 104–106, 111–113, 116, 161, 163, 167, 181, 182
photomechanical effect 86, 181, 182, 184, 189, 198, 207
photonic applications 11, 80, 83, 100, 134, 136, 145, 146, 149, 231
photoresponsive LCs 13, 64, 65, 83, 91, 125, 141, 143
photoresponsive mesogens 86, 133, 144, 159, 175, 180, 181, 184, 189

polarization 52, 53, 61, 74, 77, 78, 83, 89, 110–114, 128, 130, 153, 192, 246
  bistable 112
  nonlinear 118
polarizing optical microscopy (POM) 3, 5, 18, 21, 24, 25, 27, 93, 96, 107, 142, 155, 185, 227
poly(ethylene oxide) nanocylinders (PEO nanocylinders) 214, 232, 234, 235, 237–240, 248
polyimides 42, 43, 46, 52–54, 56, 58, 60, 61
polymer films 52, 158, 186, 209, 210
polymers 15, 53, 55, 100–103, 129, 136, 137, 139, 140, 207, 208, 214, 216, 218, 248, 250, 256, 258
  amorphous 145, 150, 182
  chromophore-doped 145
  conjugated 149
  crosslinked 15
  graft 15
  high-molecular-weight 175
  hydrophilic 161
  linear 176
  photosensitive 54, 56
  thermoplastic 43
POM *see* polarizing optical microscopy

refractive index 20, 24, 84, 85, 103, 118, 119, 137, 140, 145, 152, 153, 157, 159, 172, 173, 246, 258
refractive-index gratings (RIGs) 153, 155, 229, 231, 245–247, 258
refractive-index modulations 155, 157, 159, 246, 247

reorientation 41, 68, 69, 71, 73, 75, 77, 79, 81, 118, 119, 139, 143
  non-uniform director 122
  optically induced director 119
  photoinduced director 120
RIGs *see* refractive-index gratings

SANS *see* small-angle neutron scattering
scattering 26, 100, 107, 108, 158, 223, 228
  dynamic 9, 10
  optical 108
  small-angle neutron 22, 28
SLM *see* spatial light modulator
small-angle neutron scattering (SANS) 22, 28
smectic phases 13, 16, 17, 25, 27, 35, 73, 74
spacer 15, 72, 73, 140, 154
  aliphatic 56
  alkylene 72
  flexible 15
  soft 92, 137, 223
  stiff 15
spatial light modulator (SLM) 100, 101
SRGs *see* surface-relief gratings
substituents 2, 31, 35, 66, 73, 92, 144, 146, 171, 172, 237
substrates 45, 48, 49, 51, 52, 57, 86, 140, 159, 189, 220, 232, 234, 240, 241, 249–251, 255
  continuous 86
  glassy 230, 246
  hard 248
  patterned 234
  plastic 44
  polyimide 63
  polyimide-coated 58
  polymer 206

silicon wafer 250
surface-treated 51
surface-relief gratings (SRGs) 44, 155, 158, 229, 231, 245, 246, 257

thermal annealing 73, 78, 232, 234, 241
thermal treatment 69, 70, 76–78, 80, 93, 95, 147, 218
thermoplastics 157
thin films 53, 68, 74, 137–139, 149, 158, 169, 227, 230, 236, 240, 244, 250, 253, 255–257
transition 22, 23, 66, 67, 69, 77, 78
  crystal-to-LC 23
  disorder–order 181
  first-order 20
  order–disorder 241
  photoinduced sol–gel 117
transmittance 57

unpolarized light 79, 81, 82, 89
UV irradiation 85, 92, 97, 103, 108, 116, 156, 161, 167, 180, 182, 189, 191, 193, 194, 197
UV light 98, 99, 141, 156, 178, 183, 188, 190, 194–196, 200, 202, 204–206, 241

wavelength 20, 25, 39, 58, 64, 70, 71, 74, 83, 89, 98, 103–105, 143, 196, 201, 204
  controllable 107
  reflective 106
  telecommunication 146

X-ray diffraction (XRD) 21, 25, 26, 28
XRD *see* X-ray diffraction

yttrium aluminium garnet (YAG) 112